Education for Sustainable Development in Foreign Language Learning

This unique volume utilizes the UNESCO Education for Sustainable Development (ESD) framework to illustrate successful integration of sustainability education in post-secondary foreign language (FL) learning.

Showcasing a variety of approaches to using content-based instruction (CBI) in college-level courses, this text valuably demonstrates how topics relating to environmental, social, and cultural dimensions of sustainability can be integrated in FL curricula. Chapters draw on case studies from colleges throughout the US and consider theoretical and practical concerns relating to models of sustainability-based teaching and learning. Chapters present examples of project-, problem-, and task-based approaches, as well as field work, debate, and reflective pedagogies to enhance students' awareness and engagement with sustainable development issues as they acquire a foreign language. Insights and recommendations apply across languages and highlight the potential contribution of FL learning to promote sustainability literacy amongst learners.

This text will benefit researchers, academics, and educators in higher education with an interest in Modern Foreign Languages, sustainability education, training, and leadership more broadly.

María J. de la Fuente is Professor of Spanish in the Department of Romance, German, and Slavic Languages and Literatures at The George Washington University, USA.

Routledge Research in Language Education

The *Routledge Research in Language Education* series provides a platform for established and emerging scholars to present their latest research and discuss key issues in Language Education. This series welcomes books on all areas of language teaching and learning, including but not limited to language education policy and politics, multilingualism, literacy, L1, L2 or foreign language acquisition, curriculum, classroom practice, pedagogy, teaching materials, and language teacher education and development. Books in the series are not limited to the discussion of the teaching and learning of English only.

Books in the series include:

Education for Sustainable Development in Foreign Language Learning
Content-Based Instruction in College-Level Curricula
Edited by María J. de la Fuente

Competency-Based Teacher Education for English as a Foreign Language
Theory, Research, and Practice
Edited by Amber Yayin Wang

Globalization and English Education in Taiwan
Curriculum, Perceptions, and Pedagogies
I-Chung Ke

Advocacy for Social and Linguistic Justice in TESOL
Nurturing Inclusivity, Equity, and Social Responsibility in English Language Teaching
Edited by Christine E. Poteau and Carter A. Winkle

For more information about the series, please visit www.routledge.com/Routledge-Research-in-Language-Education/book-series/RRLE

Education for Sustainable Development in Foreign Language Learning

Content-Based Instruction in College-Level Curricula

First Edition

Edited by
María J. de la Fuente

Routledge
Taylor & Francis Group
NEW YORK AND LONDON

First published 2022
by Routledge
605 Third Avenue, New York, NY 10158

and by Routledge
2 Park Square, Milton Park, Abingdon, Oxon, OX14 4RN

Routledge is an imprint of the Taylor & Francis Group, an informa business

Library of Congress Cataloging-in-Publication Data
Names: Fuente, María J. de la (María José de la), editor.
Title: Education for sustainable development in foreign language
learning: content-based instruction in college-level curricula /
edited by María J. de la Fuente.
Description: First Edition. | New York: Routledge, 2022. |
Series: Routledge Research in Language Education | Includes
bibliographical references and index.
Identifiers: LCCN 2021026006 | ISBN 9780367530327
(Hardback) | ISBN 9781032131108 (Paperback) | ISBN
9781003080183 (eBook) Subjects: LCSH: Environmental
education—Study and teaching (Higher)—United States. |
Language and languages—Study and teaching (Higher)—
United States. | Interdisciplinary approach in education.
Classification: LCC GE70 .E3985 2022 |
DDC 363.70071—dc23
LC record available at https://lccn.loc.gov/2021026006

ISBN: 978-0-367-53032-7 (hbk)
ISBN: 978-1-032-13110-8 (pbk)
ISBN: 978-1-003-08018-3 (ebk)

DOI: 10.4324/9781003080183

Typeset in Baskerville
by codeMantra

I dedicate this book to my son Nico and my dog Teddy, my beloved daily companions while I completed this project.

Contents

Figures

Map

Tables

About the Contributors

Laura Barbas-Rhoden is Professor of Modern Languages, Literatures, and Cultures at Wofford College. She has published widely about teaching and learning in environmental humanities; global and civic learning in higher education; and ecocriticism. She has served the Association for the Study of Literature and the Envi++++ronment (ASLE) as both Co-President (2020–2021) and Co-Diversity Officer (2017–2019). She is also the author of two monographs: *Writing Women in Central America* (Ohio UP, 2003) and *Ecological Imaginations in Latin American Fiction* (UP of Florida, 2011).

Bradford S. Barrett is International Program Officer at the Air Force Office of Scientific Research and was previously a Professor of Meteorology at the US Naval Academy. He was the recipient of a Fulbright-García Robles award in 2015, and in 2018 he won the Class of 1951 Faculty Award for Excellence in Research. His educational research focuses on the ways students learn in multidisciplinary settings. He also investigates weather and climate variability across multiple temporal and spatial scales, focusing particularly on the Americas. His work has been supported by the National Science Foundation and the Office of Naval Research.

Alec Cattell is Associate Director of the Teaching, Learning, and Professional Development Center at Texas Tech University. He has published work on curriculum design and the use of collaborative tasks to foster language awareness and engagement. His literary scholarship examines representations, rhetorics, and aesthetics of disability. He spearheaded the Global Readiness Through Language and Culture project at Texas Tech University, which fostered curricular innovations across multiple language programs to support sustainability education.

Nobuko Chikamatsu is Associate Professor in the Department of Modern Languages and Director of the Japanese Language and Studies Program at DePaul University. She teaches Japanese language, Japanese Linguistics, and translation. Her research interests include second language acquisition, Japanese pedagogy, translation, and Japanese-American history and literature. Her articles have appeared in *Foreign Language Annals*, *Japanese Language Education around the Globe*, *The Modern Language Journal*, and

Studies in Second Language Acquisition. She holds a PhD in Linguistics from the University of Illinois Urbana-Champaign.

Sharika D. Crawford is Professor of History at the US Naval Academy and the recipient of the Admiral Jay L. Johnson Professorship in Leadership and Ethics. She received the Apgar Teaching in Excellence Award in 2014. Her research focuses on Latin American and Caribbean history and the Scholarship of Teaching and Learning in World History education. She is the author of the book *The Last Turtlemen of the Caribbean: Waterscapes of Labor, Conservation, and Boundary Making.* She is the recipient of prestigious grants and fellowships from the National Endowment of the Humanities, the Fulbright Scholar Program, and the American Philosophical Society.

María J. de la Fuente is Professor of Spanish at the George Washington University. Her area of research is Second Language Acquisition (SLA), in particular the effect of pedagogical tasks and focus on form, and the role of the first language in classroom second language learning. The results of her research have appeared in refereed journals such as *Studies in Second Language Acquisition* and *Language Teaching Research.* She is the author of *Gente: a task-based approach to learning Spanish* (Pearson), and *Puntos de Encuentro: a cross-cultural approach to advanced Spanish* (Cognella). Her current pedagogical research focuses on sustainable development education through content-based language pedagogy.

Megan M. Ferry is Professor of Chinese and Asian Studies in the Modern Languages and Literatures Department at Union College. She holds a PhD and MA from Washington University–St. Louis in Comparative Literature. She is the author of *Chinese Women Writers and Modern Print Culture* (2018, Cambria) and has published on gender, media, sexuality, and China-Latin America/China-Africa relations in *Modern Language Quarterly, Modern Chinese Literature and Culture, Journal of Contemporary China,* and *Continuum,* in addition to writing the middle school Chinese curriculum for the City of Schenectady, NY Public Schools.

Margaret Gonglewski is Associate Professor of German and International Affairs at George Washington University, where she directs the German language program. Her research has centered on critical issues in second language pedagogy, including the role of technology in language teaching and learning, multiple literacies, and peer-assisted learning. She co-authored the introductory German textbook *Treffpunkt Deutsch* (Pearson). As coordinator of Business Language Programs with GW's Center for International Business Education and Research, she regularly partners with Anna Helm; their scholarly collaboration has resulted in many teacher-education seminars and articles on business language education. She is currently editor of the *Global Business Languages* journal.

Anna Helm is Associate Teaching Professor of International Business at the George Washington University, where she serves as faculty director of the

Center for International Business Education and Research (GW-CIBER). Her book *The Intersection of Material and Poetic Economy: Gustav Freytag's Soll und Haben and Adalbert Stifter's Der Nachsommer* appeared in 2009. She is currently pursuing research on cross-cultural differences in consumer perceptions of green products, sustainability innovation, and curricular best practices in International Business. She also has a long-standing research agenda in Business Language Studies with Margaret Gonglewski, focusing on business case methodology, sustainability pedagogies, and real-client consulting projects.

Richard Kern is Professor of French and Director of the Berkeley Language Center at the University of California at Berkeley. He teaches courses in French linguistics, language, and foreign language pedagogy, and supervises graduate teaching assistants. His research interests include language acquisition, literacy, and relationships between language and technology. His previous publications include *Screens and Scenes: Online Multimodal Communication and Intercultural Encounters* (2018), co-edited with Christine Develotte; *Language, Literacy, and Technology* (2015); *Literacy and Language Teaching* (2000); and *Network-Based Language Learning: Concepts and Practice* (2000), co-edited with Mark Warschauer.

Belinda Kleinhans is Associate Professor of German at Texas Tech University. She contributes to teacher education by co-shaping and participating in grant-funded projects related to sustainable education and training of transferable skills in German language and culture courses. Her study on the pedagogical aspects of the topics of persecution and displacement through literature was published in *Flucht-Literatur. Texte für den Unterricht. Schneider Verlag* (2017) by D. Wrobel & J. Mikota. Her research also explores biopolitics in literature, cultural and literary animal studies, and issues of representation regarding memory, violence, and social justice in graphic novels.

Yan Liu is Assistant Professor of the Practice in the Department of Asian and Middle Eastern Studies at Duke University. She holds a PhD in Second Language Acquisition from Carnegie Mellon University. Her general research interest lies in teaching and learning Chinese as a second, foreign, or heritage language. Her recent research focuses on writing in Chinese as a heritage language, service-learning, innovative teaching pedagogies (like teaching Chinese through arts and games), Cultures and Languages Across the Curriculum (CLAC), and interdisciplinary Chinese curriculum development.

Jorge Méndez Seijas is Preceptor of Spanish in the Department of Romance Language and Literatures at Harvard University. He holds a PhD in Spanish Linguistics from Georgetown University. His research interests include second language acquisition and pedagogy, curriculum design, and phonetics and phonology. His work in curriculum design has included courses for Spanish as a second language and Spanish a heritage language. He received the 2017–2018 Graduate Global Citizen Award from Georgetown University, and the 2020 AAUSC Innovation in Language Program Direction Award: Racial/social justice.

María Luisa Parra is Senior Preceptor in the Department of Romance Languages and Literatures at Harvard University. She has a PhD in Hispanics Linguistics from El Colegio de México. Her research focuses on the pedagogy of Spanish as a heritage language, identity theory, and U.S.-Mexico transnational education. She directs the department's initiative on Spanish as heritage language, and partners with Harvard art museums and non-profit organizations to provide students with transformative learning experiences. Babson College recognized her work with the 2019 Most Innovative Professors Award. Her latest publication is *Enseñanza del español y juventud latina* (2021, Arco Libros).

Silvia M. Peart is Professor of Spanish and Second Language Acquisition. She is the recipient of the 2020 and 2021 Admiral Jay L. Johnson Professorship in Leadership and Ethics. She was awarded the 2016 Excellence in Service Award from the Research Special Interest Group at the American Council on the Teaching of Foreign Languages. She is the recipient of the prestigious Minerva and Office of Naval Research Grants, and the United States Naval Academy has recognized her research with the Ernst Volgenau Fellowship to support her investigations on language and immigration.

Deborah Reisinger is Associate Professor of the Practice of French and Associate Director of the Markets and Management Studies Program at Duke University. As Director of Duke's Language Outreach Initiatives, she establishes connections between language proficiency and the disciplines. She is lead author of *Affaires globales: S'engager dans le monde professionnel en francais, niveau avancé* (Georgetown U. Press, 2021), a textbook for students' use of language skills in professional environments, and co-author of *Community-based Language Learning* (2019), a methodology of critical service-learning in world languages. She has authored numerous articles on language pedagogy, French for the Professions, and intercultural competence.

Vesna Rodic is Lecturer in French and the coordinator of the second-year French language program at the University of California at Berkeley. She teaches language, literature, and foreign language pedagogy courses. Her research interests include second language acquisition, second language writing, as well as the relationship between aesthetics and politics in modernist literature. Her previous publications include *The New Nationalism and the First World War* (2015), co-edited with Lawrence Rosenthal.

Yerko Sepúlveda is a PhD Candidate in Spanish Linguistics at Texas Tech University, where he teaches undergraduate courses in the Spanish

Foundations Program. He holds an MA in Applied Linguistics and an MEd in Bilingual Education. He currently works as an instructional coach for the online course 'Thinking and Learning in the Maker-Centered Classroom' at Harvard Graduate School of Education and teaches Spanish at Hawken School. Prior to this, he was director of Teaching & Learning at the Technological University of La Serena, Chile. His research interests include 21st-century teaching frameworks, participatory creativity, and critical pedagogy.

Kristen M. Turpin is Senior Lecturer of Hispanic Studies at Brandeis University, where she teaches all levels of language, culture, and literature classes. Her research on second language teaching and learning focuses on classroom practice and curriculum design, primarily in the contexts of multiliteracies pedagogy, content-based instruction, and proficiency-oriented instruction. Her scholarly work in 19th-century Latin American literature has appeared in *Hispanófila and Revista de Estudios Hispánicos*. More recently, her research on a multiliteracies approach to peer review was published in *L2 Journal*.

Sandra Valnes Quammen is Senior Lecturer at Duke University. She is the director of the French Language Program, where she oversees curriculum development and implementation for first through fifth semester French language courses. Her research focuses on the intersections between language education and sustainability. She also researches multiliteracies pedagogy, especially at the beginning and intermediate levels, as well as multilingual and translanguaging pedagogies in the L2 classroom.

Raychel Vasseur is Assistant Professor of Spanish and the Director of Spanish Foundations at Texas Tech University. She holds a PhD from the University of Iowa (2018). She teaches graduate and undergraduate courses in second language pedagogy, applied and Hispanic linguistics. Her research interests include language acquisition in the study abroad context, willingness to communicate, L2 multiliteracies instruction, the development and assessment of intercultural (communicative) competence, computer-assisted language learning, program evaluation and outcomes assessment, and second language speaking skills.

Edgar Virgüez completed his PhD in Environmental Sciences and Policy ('21) and an MSc in Environment ('18) at Duke University, as well as an MSc in Environmental Engineering ('10) at Universidad de los Andes (Colombia). He is an energy systems engineer promoting the transition towards a decarbonized electric power system. As a Latino with a commitment to his ethnic and racial roots, he has also a passion for improving the higher education sector.

Acknowledgements

First and foremost, I am in debt to Charlotte Melin, at the University of Minnesota, who gave me the chance to contribute to her pioneering volume *Foreign Language Teaching and the Environment*. Her work was an inspiration to me. I thank her for also encouraging me to pursue this project.

I am also deeply grateful to Elsbeth Wright, Editor of Routledge Education. From the moment she saw the project she enthusiastically supported it, always giving me constructive feedback and encouragement. I am equally grateful to AnnaMary Goodall, Editorial Assistant of Routledge Education, for her impeccable work guiding me through all details of my first editorial authoring job. Thank you both!

Introduction

Toward Education for Sustainable Development (ESD) in Foreign Languages

María J. de la Fuente

What Is ESD, and Why Should It Be Integrated into College Foreign Language Curricula?

We face a planetary emergency, the most obvious problems of which are the degradation of ecosystems, the loss of biodiversity, the depletion and destruction of essential resources, pollution of all kinds, and extreme poverty for billions of human beings. These problems have been caused mainly by lack of knowledge—the Earth's resources are unlimited, climate change is a hoax, this whole thing is about the environment...—therefore, they are primarily an education problem. How can we transform our educational systems so the next generations can live sustainably but also recognize their responsibility in ensuring the continuity of sustainable economic, social, and environmental systems for future generations?

One of the most significant attempts to address the complex global challenges of our time is the United Nations' 2030 Sustainable Development Agenda, with its 17 Sustainable Development Goals (SDGs). UNESCO considers education one of the central elements of sustainable development. This makes sense, considering that we all need to *learn* how to live sustainably. Their Education for Sustainable Development (ESD) framework, a guide for institutions and educators to redesign curricula and pedagogies around sustainability principles, is used by hundreds of universities worldwide. Another central document to ESD is the Education 2030 Framework for Action[1] (FFA), which explicitly addresses SDG 4: "Ensuring quality, inclusive, and equitable education for everyone." One of SDG 4's seven areas of action states, "By 2030, ensure that all learners *acquire the knowledge and skills needed to promote sustainable development* [...]" (UNESCO, 2015, p. 21, emphasis added).

The concept of sustainability—meeting the needs of the present without compromising the ability of future generations to meet their needs—is present today in most areas of knowledge to the extent that it is already recognized as a discipline within the educational sciences. In many areas of the world, ESD has acquired a political and formal legitimacy, so much so that governments are starting to establish it as required content in their educational systems.[2] Concerning higher education, there is a consensus regarding the role of universities in promoting education and action toward sustainable development.

DOI: 10.4324/9781003080183-1

Besides UNESCO, many leading academic, cultural, and political organizations advocate a global approach to teaching and learning about sustainability across disciplines, urging countries to integrate education for sustainable development in higher education contexts.

The reality is that the global sustainability agenda is shaping most higher educational institutions worldwide. Many are already implementing policies and promoting curricular initiatives to tackle social and environmental sustainability. According to Leal Filho (2010), the university cannot excuse itself from meeting one of the most significant challenges facing humanity. Filho argues that, although important at all educational levels, ESD is most crucial at the university stage because students will soon pursue careers in diverse professions, and they will need to know what impact their professions can have in solving sustainability problems. ESD will motivate them "to take action both during their time as students and, later on, as professionals" (Filho, 2010, p. 2). Therefore, undergraduates should acquire, through innovative teaching and learning, competence-based sustainability awareness and literacy, so that they can solve the challenges they will encounter in their professional fields.

A consensus also exists that, to spread education and culture about sustainability, higher education cannot limit it to creating majors and minors in sustainability, or graduate degrees in Environmental Studies, as important as these are. ESD should permeate the curricula of all the university's disciplines and areas of knowledge, something that, given the area's multidisciplinary and multidimensional nature, seems just logical. There are at least three dimensions of sustainability: *environmental, social,* and *economic,* which interact in complex, systemic ways. A fourth *cultural* dimension has been recently added that underscores the need to maintain cultural practices and heritage for future generations (Payne & O'Neil, 2019). As social, economic, environmental, and cultural themes, problems, ideas, discussions, and debates start permeating the curricula across universities, these may soon realize that their old departmental, disciplinary divisions no longer correspond to reality because the solution to a sustainability problem cannot be found in one single department, not even in one single school.

College foreign language departments and programs could have an essential part in ESD. Sustainable development interdisciplinarity is not limited to the sciences, nor even the social sciences. Fields such as history, ethics, philosophy, anthropology, literature, religion, art, cultural studies, education, second language acquisition, or foreign language pedagogy add crucial perspectives (intellectual, moral, global, cultural and cross-cultural, linguistic, and cross-linguistic) to the sustainable development debate.

Foreign Language Education as ESD

This volume's position is that foreign language education could and should acquire an essential role in sustainability literacy within the higher education landscape. Because of the scope of disciplines usually found inside such foreign language (FL) departments, environmental, social, and cultural dimensions of sustainability can be addressed from various perspectives. But just like the university's fragmented structure needs to evolve, FL departments need to do the same (Lomicka & Lord, 2018).[3] Just as the key to sustainability lies in balancing this complex system where economic, social, and environmental needs interact in multiple and complex ways, engaging in sustainable education at a curricular level would entail engaging in systemic and relational thinking with different areas, subjects, categories, and structures *within* departments. Furthermore, it would require abandoning existing fragmentations and disconnections (language and literature, language and content, lower-level courses and higher-level courses, tenure track and non-tenure track faculty, research, and teaching). It would also mean something even more difficult but more important, *"Hacer un esfuerzo por aprender a desaprender"*—making an effort to learn to unlearn—(Prádanos, 2015, cf. Halberstam, 2012), in other words, connecting our disciplines to others in a systemic way that could benefit our students and our society. Prádanos (2015), in a pioneering article where he made a case for the integration of sustainability in all levels of college Spanish classes, offered two options: Spanish programs could obey to "inertia" or instead set up "diverse, interdisciplinary, critical and holistic programs that lead the necessary paradigm shift in higher education towards a sustainable cultural model, with global awareness, fair to the entire biotic community and socially desirable" (Prádanos, 2015, p. 343). Prádanos' advice is good for all FL programs. What prevents foreign language departments from being relevant in the sustainability discourse that will surely frame higher educational agendas in the 21st century?

Transforming Foreign Language Departments

One would think the choice is easy, but in the field of FLs, which fares "rather poorly when it comes to change" (Martel, 2017, p. 431), things are not that simple. Let us explore what seems to be the best of the two options: That we decide to do away with fragmentations and disconnections such as the 'language-literature' divide which, under an interdisciplinary model, is counterintuitive because today's FL departments gather many disciplines besides those two. In fact, we should agree with Pfeiffer (2008) that "The disciplinary anchor to FL departments' teaching and research mission is its particular language because that is the one *distinctive object of study*" (p. 298). More than 20 years ago Janet Swaffar, in an essay published in *MLA Profession* titled "The case for foreign languages as a discipline," had already described what the mission of such departments would be, "Our discipline's goal, therefore, is to enable students to do things with words and to recover what has been done with words, socially, historically, politically, and interpersonally. Each subfield does so in different ways" (Swaffar, 1999, p. 158). In today's context, we should expand this definition as follows: to recover, study, or analyze what has been done or is being done with the specific language—culturally, literarily, pedagogically, socially, historically, politically, interpersonally, etc., depending on faculty's academic interests. This would include making connections of various types: By relating it to local and/or global realities, exploring it from multiple perspectives, and/or—as is the case of the authors included in this volume—connecting it to sustainable development education. Along with this, the educational mission would be to make this disciplinary and interdisciplinary knowledge available to our students while enabling them to become multicompetent users of the target language, so they can contribute to a sustainable future.

This last point assumes that, in this new configuration, another related disconnection would no longer exist: the 'language-content' divide. Instead, there would be a four-year, coherent, content-based FL curricula in place where faculty would teach multidisciplinary courses in the target language (TL), and the TL would be both the medium and object of study. This is the nature of content-based language instruction (CBI), arguably the most effective framework for transforming college FL education.[4] As mentioned, in this scenario there is no 'language' faculty or 'literature' faculty because *the target language* is our disciplinary field and, thus, language acquisition is the main goal (among others) in every class of the curriculum and, therefore, all faculty are language faculty (in addition to their other disciplinary fields). With this vision in mind, there is work to do at the curricular and instructional levels. Depending on the specific institutional context, this requires redesigning courses and/or creating new ones while ensuring articulation between them and across other university units. Pedagogically, the curriculum needs to be informed by constructivist approaches to learning—pedagogies like problem-solving or community engagement that emphasize collaborative, participatory learning, balance content-language instruction, and get learners to their highest possible levels of linguistic development.

But things get complicated by two facts that the profession has grappled with for a long time: one is the "well-known undervalued view of language pedagogy in foreign language departments, where language teaching is a task, and the goal of many faculty members is *not* to teach language courses" (Pfeiffer, 2008, p. 298); and the other is that the integration of an informed approach to second language (L2) development into upper-level courses has not been a priority. This second fact is a consequence of the first one, which, in turn, is a consequence of the nature of graduate programs in FLs, which lack an adequate research-based educational component on second language acquisition, curricular, instructional, and assessment practices. This fundamental inequality in the value of the 'language' discipline in relation to literature (or humanities, generally speaking) is at the core of the fragmentary status of FL departments.

At the beginning of this section, we asked what it would take to adopt this systemic form of FL curriculum with a learner-centered educational vision. The answer is clear: It would take a significant change but, as Martel (2017) states, "maybe an upheaval is what it will take to ensure a consistently valuable education for our country's foreign language learners" (Martel, 2017, p. 431).

Why ESD in Foreign Languages?

The presence of environmental and social sustainability topics in FL courses offered on US college campuses, and the current scholarship and courses taught in the field of the environmental humanities, speak to growing interest in this area. Eleven of the essays in Melin's (2019) volume *Foreign Language Teaching and the Environment* represented concrete examples of upper-level FL courses or curricular initiatives that promoted environmental literacy as a desired competency that "can become a starting point for broader transformation and curricular coherence" (p. 8). Many of these are linked to ecocriticism—part of the environmental humanities—and use literary texts as primary materials; others utilize various forms of cultural representations and are framed under content-based language instructional approaches. For the most part, these are isolated initiatives—they do not constitute curricular integrations but individual courses—and thus remind us of the fragmented nature of FL departments. Nevertheless, Melin's (2019) pioneering volume intended to start a conversation, and it certainly did.

In language programs, the movement to CBI and active learning approaches to teaching and learning FLs (such task task-based language teaching or community-engaged language learning) naturally gravitates to sustainability. Content-wise, it is of high interest among educators and students; it covers a wide variety of areas and issues, content is readily available in various genres and media formats, multiple connections and perspectives can be established, and it involves various disciplines. In addition, there is a natural fit between the pedagogy of ESD and FL instructional practices that have been identified through research as effective. In general, ESD favors learner-centered instructional approaches that enrich the learning process. Project-based learning, community-engaged learning, task-based, and problem-based teaching and learning are all as much a part of sustainability education. Likewise, ESD teaching is characterized by being contextualized, incorporating the study and treatment of local and global problems, and reinforcing collaboration with local entities. This lends itself to community-engaged pedagogies in FL education and natural connections to the communities and cultures where the TL is used.

The most apparent connection between ESD and FL learning is the fact that multilingualism and multiculturalism are part of our present reality. Therefore, monolingualism and monoculturalism may be an obstacle to sustainability efforts at the global level. The abilities to understand and tolerate culturally different points of view, negotiate differences, and develop sensitivity to local cultures, are vital to sustainability negotiating efforts. The following would be an example: Since sustainability education requires the development of problem-solving and negotiation skills, reinforcing students' communicative competencies (sustainability-specific) in FLs is vital for what future professionals will need in their work. This is a valid example that establishes a connection at the language level—potentially bringing many subdisciplines from Linguistics into the mix—and explores important cultural differences. However, this is only one of the hundreds, potentially thousands of ways to connect our disciplines to ESD.

The title of this section is *Why ESD in Foreign Languages?* Joining Filho's sentiment, I believe we cannot excuse ourselves from the challenge of sustainable development. As educators, we have a position that allows us to educate young people who will enter the job market and try to be a part of the solution. FL departments could contribute to that in significant ways. But the reader will find more than a dozen other good reasons in the papers included in this volume.

The Volume: Education for Sustainable Development in Foreign Language Learning

This volume was conceived with three main goals: The first one was bringing together a collection of papers that would illustrate and bring forward the relevance of FL education in the sustainability discourse. This would include analysis of theoretical and pedagogical approaches to develop sustainability-oriented curricula, as well as state-of-the-art interdisciplinary curricular innovations in the field of FL education that constitute effective models of CBI that incorporate sustainability learning goals for their students. Embedded in this goal—under the word 'effective'—is the fact that the applied research papers included in this volume describe course-level or sectional-level[5] curricular projects that

a   are well-grounded in Second Language Acquisition (SLA) research and/or FL pedagogy
b   have an interdisciplinary approach and a connection to sustainable development
c   constitute unique cross-disciplinary approaches to sustainability that only FL study can provide
d   make knowledge available to students while they become multicompetent users of the target language (i.e., via CBI)
e   use learner-centered pedagogies (collaborative, participatory learning)
f   promote critical, higher-order thinking skills, ESD-specific, and other competencies
g   integrate attention to language development (upper-level courses)
h   seek highest possible levels of linguistic development (upper-level courses)

The papers included in this volume lay the theoretical, pedagogical, and conceptual basis of what ESD in FLs is today but also of what it *could be*, and for this reason, they constitute invaluable work. Although these projects were carried out as units within courses, or individual courses, or sections of a curriculum—thus, still pieces of the four-year program puzzle—they are informed projects assessed with educational research methods examining students' experiences, perceptions, and outcomes. Thus, they can impact the design, delivery, and assessment of similar projects in other departments. Additionally, all authors situated their work within the broader discussion of the status of FLs in higher education, emphasizing the fundamental role that FL education can have in ESD.

Some papers represent individual essays and projects whereas others report on projects developed as intradepartmental or interdepartmental faculty collaborations. The 12 papers in the volume were written by a total of 22 authors. Most (17) were faculty in FL departments (two in Chinese, four in French, three in German, one in Japanese, and seven in Spanish), and four were faculty from other disciplines—Business, Environmental sciences, History, and Meteorology—collaborating in interdisciplinary projects. Eleven FL faculty members had SLA or FL Education backgrounds, while six had Literature/ Humanities backgrounds. It is of relevance to this volume to underscore that, in each of these projects, every faculty member (a) had the relevant knowledge base to undertake it, and/or took upon themselves to acquire it, or (b) when working as a team, deferred to the expertise of a faculty member who had that knowledge base (sustainability discipline, second language learning research, pedagogy, etc.). This ensured that the needed expertise was there in terms of content and language development and, for this reason, these papers meet the criteria outlined above. The sustainability-specific goals varied in scope, approach, level, and scale of integration, depending on the specific context of each project.

Between Applied and Action Educational Research in Foreign Languages

The second goal of this volume was to adopt an educational research perspective to course/curriculum development in FLs. This involved a detailed description and justification of course design, learning objectives—including ESD-related goals—materials, activities, and instructional approaches, among other aspects. It also meant gathering data from students as a regular part of their activities, tasks, and projects for assessment purposes, some of which could eventually be used to assess aspects of the project. Most used students' data from surveys, questionnaires, focus groups, written reflective thinking protocols, or interviews to gather feedback on disposition, feelings, and opinions, gathering insightful and detailed information which they analyzed to get insights on awareness, abilities, perceptions, language, sustainability-related goals, etc. Thus, these papers do not constitute essays as much as interdisciplinary projects undertaken from an educational applied research perspective.

Because of the experimental nature of these projects, they are usually concerned with the effect of the practice (in this case, the course or pedagogical unit implemented) in specific educational settings. For this reason, most of the papers take an *action inquiry*, exploratory, and reflective approach to describing and understanding how the pedagogies, the materials, or other aspects of the course affected their own students in their own classes. This, however, does not mean that the project was not the result of an interest in solving a university-wide or a profession-wide problem. Furthermore, they often point to the limitations and challenges of their projects and offer recommendations. In sum, these eight projects constitute excellent examples of *educational action research* in FLs and, in the case of the last four, *participatory action research*. However, a solid theoretical background, clear goals, and the will to advance the foreign language education discipline, would place them closer to applied research methodology.

New Pedagogies for ESD and FL Education

This volume's third and final goal was to underscore the importance of pedagogical frameworks of a constructivist and *student-centered* nature for a 21st-century education that supports sustainability-based FL instruction. As mentioned earlier, both ESD and FL education share the common belief that learning is a self-directed, sociocultural process (at least to some extent, depending on the specific learning approach), so students have an important role in their learning process. They also share a preference for *collaborative* learning, valued as a competency in itself (i.e., problem-solving) and as a way to develop the TL, respectively. The first four papers in the volume discuss some of these frameworks and specific pedagogies from a theoretical and educational perspective, using examples from actual courses. The following eight papers report on curricular projects that used various learner-centered pedagogical approaches. Needless to say, they all reside under the 'umbrella' of CBI. These are the instructional approaches[6] illustrated in this volume:

- inquiry-based learning (Chapters 7, 10)
- experiential learning (Chapters 1, 7)
- multiliteracies pedagogy (Chapters 2, 5)
- project-based language learning (Chapters 3, 5, 6, 7, 8, 11)
- Problem-based pedagogy and case pedagogy (Chapters 4, 9, 11)
- community-engaged learning (Chapters 1, 7, 11)
- critical content-based instruction (Chapters 5, 8, 10)
- task-based language learning (Chapter 4)
- translanguaging pedagogy (Chapter 12)

The variety of approaches showcased in this volume also lets us foresee the types of rich, long-lasting learning experiences that all FL university students—not just those taking part in these courses—could have and would want to have.

Organization of the Volume

This volume is divided into three parts. **Part 1** introduces the reader to theoretical and pedagogical approaches that can be used to develop interdisciplinary—sustainability-oriented—content-based courses at all levels of FL curricula. Laura Barbas-Rhoden's paper (**Chapter 1**) was placed strategically to open the book for two main reasons: first, it establishes the link between Melin's (2019) volume *Foreign Language Teaching and the Environment* and this one, by bringing concepts from the environmental humanities field that are relevant to sustainability and connecting then to CBI; second, it allows us to see *what is possible*, by masterfully connecting environmental humanities content to advanced Spanish language learning using community-engaged pedagogy. Barbas-Rhoden is a 'must-read' for departmental colleagues outside of the humanities, such as faculty in SLA or FL education, because it enriches and expands our view of ESD and is grounded on curricular and pedagogical knowledge. In **Chapter 2**, Kristen Turpin first introduces the readers to the *multiliteracies pedagogy* framework. Then she uses an instructional unit on ecotourism as a theory-to-practice approach to show how this pedagogy unfolds in a real class and how learners can develop sustainability literacies. In discussing this pedagogy's emphasis on exposing learners to multiple texts and genres, the author underscores the need for course design to "prepare students to participate in real-world discourse communities committed to sustainability." Turpin does so herself by integrating a multiliteracies unit on environmental conservation into her own course and skillfully inserting a literary work. The study expands a knowledge base of theory-to-practice strategies that span both literary and pedagogical disciplines. In **Chapter 3**, Megan Ferry gives us a detailed account of *project-based language learning* (PBLL). She proposes PBLL as a curricular framework for deep learning about human and ecological sustainability. Because it is a form of self-directed learning, Ferry says that it facilitates "student intellectual and emotional growth." The brief account she offers of a PBLL Chinese course on climate action is a 'textbook example' of the type of ESD that only an FL course can provide: "Students learn that other cultures see their relationship to nature and the environment differently. The language they use to speak about these relationships can open many perspectives for seeking solutions to our current climate and humanitarian crises." In the last chapter in this section, **Chapter 4**, María J. de la Fuente presents *problem-based learning* (PBL) as an optimal instructional approach for courses or curricula focusing on the development of advanced language abilities and competencies related to ESD. In the chapter, de la Fuente explains how, through working with sustainability cases, students develop cross-cultural sustainability knowledge and competencies. Furthermore, working with cases can improve their language abilities and make them more effective multicompetent language users. The author mentions that developing ESD-based interdisciplinary courses using learner-centered pedagogies requires a broad knowledge base, citing this as an opportunity to explore new disciplines and develop new forms of teaching.

Part 2 includes four papers that illustrate current best practices in sustainability-based curricular initiatives, assessing them from an applied research perspective. In **Chapter 5**, Jorge Méndez-Seijas and María Luisa Parra explain how they redesigned an Elementary Spanish course using the *multiliteracies framework* with the goal of engaging their students "with social, cultural, and environmental sustainability issues faced by local, national, and global communities." They used surveys and interviews to obtain reflective information from students regarding their perceptions on achieving learning outcomes. The positive results evidenced by the results of these surveys are the reason Méndez-Seijas and Parra underscore the relevance of FL literacy from the beginning of instruction to create "a socially and environmentally responsible glocal citizenry for the 21st century and beyond." **Chapter 6** describes Raychel Vasseur and Yerko Sepúlveda's design, implementation, and assessment of The Migration Project, an instructional unit for a fourth-semester Spanish course that contains high-impact instructional strategies like thinking routines and guided reflections. The unit has SDG 4.7 as a learning goal, impacting approximately 1,000 students per year. Goal 4.7 states that "by 2030, all learners would acquire the knowledge and skills needed to promote sustainable development, including human rights, gender equality, and the appreciation of cultural diversity and a culture's contribution to sustainable development." Vasseur and Sepúlveda use students' essays as data to analyze students' acquisition of specific "dispositions" such as "taking action toward inclusive and sustainable societies" or "understand perspectives," finding that students were able to demonstrate them, which proves that a "sustainability-based, content-enriched pedagogy" can be applied from the initial courses of the curriculum. In **Chapter 7**, Richard Kern and Vesna Rodic describe a sustainability-based pedagogical unit created for an Intermediate French course that focuses on France's *La Charte de l'environnement,* where students analyze primary sources, explore French attitudes about environmental issues, and perform debates. The authors attribute the success of this course to the use of three learner-centered pedagogies: Inquiry learning, experiential learning, and the creation of discovery pathways and research connections. Using an action research approach, Kern and Rodic examined students' production of materials and responses to questionnaires. Their findings suggest that "research skills were improved, students' pathways of exploring French culture were individualized, and their engagement with environmental sustainability was expanded through exposure to historical, literary, cultural, and political perspectives." **Chapter 8** is Alec Cattell and Belinda Kleinhans' applied research paper on the implementation and assessment of a curricular project: the redesigning of three German courses, using Critical CBI and PBLL pedagogies. The authors collected data from two sources—assessment rubrics and surveys—and among the insights obtained from it was the fact that all students reported development of L2 proficiency "while acquiring additional transferable skills traditionally not associated with a German degree." The authors mention, among others, critical and creative thinking, analytical skills, or awareness of global challenges. Cattell and Kleinhans underscore the value of PBLL and CBI "not only for sustainability education but also for sustainable education."

Part 3 contains four papers illustrating best practices in interdisciplinary collaborations that integrate sustainability education in FL departments. The papers present the projects and their assessment from an applied research perspective. In **Chapter 9**, Margaret Gonglewski and Anna Helm report on a collaboration to design and implement a sustainability-based pedagogical unit for an already cross-disciplinary course: Business German. The "green business" unit centered on Germany's energy transition and used case study pedagogy, a common instructional technique in the business discipline, within the broader context of the CBI framework. Gonglewski (an FL professor) and Helm (a business professor) used an action research approach to obtain students' perceptions on sustainability-specific learning outcomes. The authors mention how students felt the course "had an impact on their awareness of sustainability as a locus of citizen activism." In **Chapter 10**, Silvia M. Peart, Bradford S. Barrett, and Sharika Crawford describe their collaboration designing and teaching a cross-disciplinary language learning module on *migration* for an advanced Spanish seminar. By having disciplinarity expertise from three different fields, the authors could bring a systemic thinking perspective to the topic exploring historical patterns of migration, the links between migration and climate change-related events, and issues of language and identity. Peart, Barret, and Crawford's detailed examination of learners' reflections, assessed with a rubric and quantitatively analyzed, indicated a very high level of engagement with the class content, high level of critical awareness, and high use of specialized, sustainability-related vocabulary. Deborah Reisinger, Sandra Valnes, Yan Liu, and Edgar Virgüez's paper (**Chapter 11**) reports on a multidisciplinary collaboration of an exploratory nature using the Cultures and Languages Across the Curriculum (CLAC) Framework. Under this model, they created three tutorials (Chinese, French, and Spanish) and aligned them with each of the four topics of an Environmental Sciences course: Environmental justice, climate change and wildfires, hydropolitics, and biodiversity. Each tutorial used several pedagogies (case studies, project-based activities, and community-engaged learning). Students reported high levels of perceived gains in language development and understanding of sustainability concepts. Finally, the authors emphasized the value of CLAC for sustainability-related, FL curriculum development. In the final chapter, **Chapter 12**, Nobuko Chikamatsu also used the CLAC framework to develop a half-credit, advanced Japanese course taught parallel to an Environmental Ethics course. The course's theme were the diseases caused in Japan by two environmental disasters. This is a learner-centered course where Chikamatsu applies experiential learning, multiliteracies, and translanguaging pedagogy, one of the main foci of Chikamatsu's paper. Using a conversation analysis approach, she analyzes students' collaborative and individual translation work (respectively) of two literary texts. She concludes that the usage of two languages, i.e., translanguaging, facilitated learners' discussion of environmental justice topics and their understanding of institutional dichotomies, such as state versus people or city versus village, which are recurrent elements of environmental Japanese literary work.

Notes

1 UNESCO and several other organizations signed this Declaration after the World Education Forum 2015 in Incheon, S. Korea. Participants from 160 countries adopted the Declaration.
2 https://www.nytimes.com/2019/11/05/world/europe/italy-schools-climate-change.html Retrieved May 12, 2021.
3 According to a survey (Lomicka & Lord, 2018), 40% of FL departments' faculty and administrators think that the two-tiered system still exists.
4 It is an area in expansion in Second Language Acquisition (SLA) research (Larsen-Freeman, 2018), and its implementation in language education is growing (Troyan et al., 2017).
5 It refers to portions of a curriculum, such as the second year of the 'language curriculum.'
6 It is necessary to point out that some overlap exists between these categories, since some are broader than others. For example, problem-based and case-based pedagogy are both examples of inquiry-based learning; and community-engaged learning is also a type of experiential learning.

References

Halberstam, J. (2012). Unlearning. *Profession*, 9–16. Nueva York. Retrieved August 17, 2021, from http://www.jstor.org/stable/41714132

Larsen-Freeman D. (2018). Looking ahead: future directions in, and future research into, second language acquisition. *Foreign Language Annals, 51*(1), 55–72.

Leal Filho, W. (2010). *Universities and climate change: Introducing climate change to university programmes.* Berlin: Springer Publishing.

Lomicka, L., & Lord, L. (2018). Ten years after the MLA report: What has changed in foreign language departments? *ADFL Bulletin, 44*(2), 116–120.

Martel, J. (2017). Is the field of foreign language education disposed to change? *The Modern Language Journal, 101*(2), 431–433. https://doi.org/10.1111/modl.12422

Melin, C. (2019). *Foreign language teaching and the environment: Theory, curricula, institutional structures.* The Modern Language Association of America.

Payne, L., & O'Neil, J. (2019). Cultural sustainability in higher education. In W. Leal Filho (Ed.), *Encyclopedia of sustainability in higher education.* Berlin: Springer Publishing. https://doi.org/10.1007/978-3-319-63951-2_109-1

Pfeiffer, P. (2008). The discipline of foreign language studies and reforming foreign language education. *The Modern Language Journal, 92*(2), 296–298. https://doi.org/10.1111/j.1540-4781.2007.00719_4.x

Prádanos, L. (2015). La enseñanza del español en la era del antropoceno: hacia la integración de la sostenibilidad en las clases de español como lengua extranjera. *Hispania, 98*(2), 333–345. https://doi.org/10.1353/hpn.2015.0056

Swaffar, J. (1999). The case for foreign languages as a discipline. *ADFL Bulletin, 30*(3), 6–12.

Troyan, F., Cammarata, L., & Martel, J. (2017). Integration PCK: modeling the knowledge(s) underlying a world language teacher's implementation of CBI. *Foreign Language Annals, 50*(2), 458–476.

United Nations Educational, Scientific and Cultural Organization. (2014). *Roadmap for implementing the Global Action Programme on Education for Sustainable Development*. Paris, France: UNESCO.

United Nations Educational, Scientific and Cultural Organization. (2015). *Education 2030. Incheon Declaration and Framework for Action for the implementation of sustainable development goal 4*. Paris, France: UNESCO.

Part 1

Approaches to Integrating Education for Sustainable Development into Foreign Language Curricula

Part I

1 Sustainability and the Pluriverse

From Environmental Humanities Theory to Content-Based Instruction in Spanish Curricula

Laura Barbas-Rhoden

Introduction

Environmental humanities is an interdisciplinary field of work that theorizes and interrogates ways of thinking, being, and doing that have shaped socio-environmental pasts and present, and which must be engaged for inclusive human futures. Environmental humanities thinking emphasizes that humans are always already enmeshed in a socioenvironmental world. It recognizes ontological diversity, that is, that human understandings and experiences of the world are heterogeneous, and notes that power dynamics shape the influence of diverse ways of knowing in the world (O'Gorman et al., 2019, pp. 445–447). Thinking with the environmental humanities, particularly in its most critical expressions, opens abundant possibilities for content-based instruction (CBI) in modern languages.

How, though, may one who is unfamiliar with scholarship in the environmental humanities begin to consider its implications for CBI? And why would doing so matter? CBI is a well-established instructional approach to language learning that integrates the target language and subject matters or content themes relevant to learners (Kennedy, 2006). This orientation toward learners presents a challenge for course designers and instructors: how does one select relevant content? The identities, experiences, and aspirations of learners are heterogeneous, and increasingly so in institutional contexts in the United States. Importantly, because the environmental humanities as a field acknowledges diversity at its deepest level of expression, it can shape participatory inclusion in the learning and teaching of languages. CBI grounded in environmental humanities creates a learning environment in which each participant can reflect upon their experiences in a dynamic socioenvironmental world. It creates the conditions by which learners can engage new material and is responsive to the needs of learners; such an orientation is a key imperative for CBI (Rodgers, 2014; Stryker & Leaver, 1989). CBI anchored in environmental humanities offers learners a framework to make sense of sustainability challenges at all levels, from the local to the planetary. Thus, those who have a particular interest in possibilities at one level or area of focus are more readily able to perceive potential connections with others with different interests or expertise.

DOI: 10.4324/9781003080183-3

In making an argument for environmental humanities approaches to content-based foreign language education, this chapter joins the call for criticality in CBI (Sato et al., 2017). Sato et al. (2017) define criticality as "a reflective critique of the imbalance of power in society, and the instigation of changes needed to emancipate and empower people" (p. 51). Environmental humanities approaches are anchored in criticality about the imperative for intertwined human and environmental futures of the socioenvironmental world in which human lives are enmeshed. This socioenvironmental world is, in fact, a pluriverse, or "world of worlds" (Escobar, 2018, p. xvi). Equitable, sustainable futures are fundamentally dependent upon not only technical solutions but also widespread understanding of the historical and structural roots of inequity and environmental degradation and a commitment to co-created and sustainable futures (Prádanos, 2018, p. 19).

To grapple with questions of inclusive sustainability for human communities in the pluriverse is to develop or deepen an awareness of the operations of power that create and replicate inequities while being embedded in their workings. It may involve setting aside the idea of sustainability in favor of other concepts, such as renewal, repair, and regeneration, in recognition that dominant economic models are unsustainable in biophysical terms and have been for some time. It means acknowledging that modernity and coloniality are intertwined and have produced institutionalized practices, technologies, and processes that have structured the world in ways that distribute environmental harms and benefits inequitably at scales from the local to the planetary. Through these practices and processes, some individuals have been made to feel the effects of environmental challenges more acutely than others and have fewer resources for mitigating their harmful effects. Frequently, these same people have also been excluded from future-making because their knowledge is marginalized by dominant groups and because they belong to groups that have been disenfranchised from political processes at multiple levels of scale.

Fostering a space in which an increasingly diverse student population can make sense of the present through CBI in modern languages represents an important opportunity for the kind of imaginative, regenerative future-making that 21st-century realities require if people are to confront injustices and co-create possibilities for repairing and regenerating the natural and social systems in which individual lives are enmeshed. In order to further equitable futures, language educators must consider ontological and epistemological concerns, including what knowledge and skills 21st-century institutions legitimize and transmit, and whether such knowledge and skills foster inclusive, life-giving futures. From a disposition of ongoing critical reflection about the historical roots of contemporary systems and cultural humility concerning what has been made marginal or silenced, it becomes more possible for educators to change the quotidian, such as the curriculum, lesson plan, or a publicly engaged project. The CBI praxis that unfolds from such reflection proffers opportunities for students and instructors to discern root causes of environmental crises; contemplate the scope and nature of needed systemic change; and think critically about what is necessary for working collaboratively for change.

In the following sections, I elucidate key theoretical insights from the environmental humanities and highlight how the field challenges legacy structures and cultures, such as disciplinary silos and 'nature' and 'society' binaries that inform divisional structures in higher education. I then articulate some assumptions about teaching and learning languages in institutional settings and why contextualizing language education is essential in furthering efforts for sustainable and inclusive socioenvironmental futures. Finally, I give concrete examples of how theory informs practice, particularly as related to course design and content selection, in a Spanish gateway course for learners of Intermediate-Mid to Advanced-Low levels of proficiency.

Theoretical Insights: Environmental Humanities for Content-Based Language Instruction

What is the environmental humanities? The authors of 'Teaching the Environmental Humanities' offer a succinct framing:

> environmental humanities (EH) is an interdisciplinary field of inquiry that brings the insights and approaches of the humanities—centered on questions of meaning, value, and ethics—to bear on some of the most pressing challenges of our time. The field is grounded in the growing recognition that diverse human understandings about, and activities in, the environment are critical factors in making sense of, and responsibly inhabiting, a dynamic more-than-human-world.
>
> (O'Gorman et al., 2019, pp. 428–429)

The notions of "diverse human understandings about, and activities in, the environment" are particularly resonant for language instructors in US higher education contexts; such ideas are often signaled with terms like 'intercultural learning' and focus on sociocultural relations (often to the exclusion of socioenvironmental relations).[1]

Multiple currents shape the contours of the environmental humanities. In exploring them further, it is possible to discern how each current can shape a curriculum and how, in combination, they can transform it. Most important for the present chapter's focus on CBI are four critical streams: *decolonial* theory, *critical race* theory and studies, *critical geography*, and *degrowth* studies.[2] Concepts and tools from these fields can help instructors design curricula and course plans that make more perceptible the differentially experienced realities of the present; invite consideration of the dynamics over time that have shaped the present; and encourage the imagination of alternative futures. To put it very briefly, critical geography contributes a spatial and scalar register, and critical race theory, a historical or temporal one, especially concerning the technology of legal structures that have shaped relationships (among people, with the land). Decolonial theory centers attention on the operations and persistence of the coloniality of power, through racial capitalism and colonialism, in the Americas and other regions of the world. Postgrowth (or degrowth) studies question the dominant growth-oriented economic paradigm that has been normalized by these historical processes and invites the imagination of radically different futures. Let us consider some key concepts from the four aforementioned (and capacious) critical theoretical fields that are particularly relevant to foreign language CBI related to sustainability.

Decolonial Theory

Decolonial theory has its intellectual roots in Latin American scholars' work and Indigenous scholars in Canada and the US.[3] It draws attention to the existence of Western supremacist ideological and territorial projects in the Americas since the beginning of the colonial/modern era. At its core, it engages with the coloniality of power, a concept elucidated by Aníbal Quijano (2000) to name the logics and legacies of European colonialism, set in motion by colonial projects but not limited to their political lifespan. It posits the inseparability of modernity and coloniality, underscores the persistence of coloniality of power in states and institutions, and asserts the existence of the pluriverse, or a world of worlds (Escobar, 2018; Jazeel, 2017). Decolonial theory provides tools to consider the construction of modernity through institutions, such as universities, academies, societies, and museums; technologies, including legal structures, by which racialized capitalism and extractivism organized people as labor and land as sites for resource extraction (Gómez-Barris, 2017, p. xvii); and operations of power at different levels of scale (Escobar, 2018).

Scholarship in Indigenous studies centers Indigenous ways of being and knowing, which exist in contrast to the ecocidal logics of extractive and settler colonialism set in motion worldwide by colonization (Davis & Todd, 2017, p. 763). Whyte (2017), for example, points out that Indigenous climate change studies "arise from memories, knowledges, histories, and experiences of oppression that differ from many of the nonindigenous scientists, environmentalists, and politicians" (p. 153) and calls for non-Indigenous researchers to "grasp richer senses of their responsibilities to work with Indigenous collaborators mutually instead of exploitatively" (p. 159). Davis and Todd (2017) likewise advocate for shifting discussion about the Anthropocene from its Eurocentric framing by moving the date marking the start of the Anthropocene to the beginning of the colonial period in the Americas, in order to name "the problem of colonialism as responsible for contemporary environmental crisis" (p. 763). As they remind, "stories we will tell about the origins of the Anthropocene implicate how we understand the relations we have with our surrounds" (p. 767). An approach to sustainability-based foreign language instruction in Spanish informed by decolonial and Indigenous theory will acknowledge linguistic and cultural heterogeneity (in the Iberian Peninsula and the Americas); examine the way conquest, colonization, and the coloniality of power have re-shaped territories and socioenvironmental worlds; and consider persistence and change, including revitalization, with regard to Indigenous knowledges, languages, and practices of inhabitation.

Critical Race Theory

Critical Race Theory (CRT) shares with decolonial theory a questioning of foundations and an activist orientation that seeks transformation and not simply understanding (Delgado & Stefancic, 2001, p. 3; Tuck & Yang, 2012, pp. 2–3). CRT, originating in legal studies and transdisciplinary critical race studies more broadly, equips students with tools to understand how everyday thoughts and actions arise from assumptions transmitted and absorbed as part of living and being educated in contexts like in the US (and many other nations) which have been shaped by racialized capitalism (Delgado & Stefancic, 2017, p. 2). Just as the coloniality of power is a key concept from decolonial theory, so is intersectionality a key one from CRT. Initially articulated by Kimberlé Crenshaw (1990), intersectionality refers to how multiple social forces converge and act to influence possibilities and outcomes for particular individuals. Intersectional approaches are increasingly prominent not just in environmental justice work but also in the climate movement and public health contexts. Concepts like that of 'intersectional identity' that acknowledge the multiple aspects of one's identity, such as ethnicity, race, gender, age, and so on, that affect both how a person understands the world around them and how others perceive them, are particularly helpful for CBI in Spanish through sustainability education. Concepts and frameworks from critical race studies allow for deepened discussions of the constructions of identity in diverse cultural settings, as well as the form anti-racist work takes in different national or regional contexts or in affinity groups that may shape a person's intersectional identity (such as political parties or religious affiliations).

Critical Geography

A key insight of relevance from the field of critical geography is that "racialized processes are always and thoroughly spatialized ones" and that "diverse axes of difference are mutually constructed and thus intertwined" (Price, 2010, p. 147). Additionally, as Price (2010) notes, "though pervasive, race is neither static nor immutable" (p. 154). Curricular materials informed by the scalar and temporal perspective from critical geography can convey a sense of history in which both continuities and ruptures become more perceptible. How students and instructors understand environmental challenges and their relationship to the systems that produce harm and challenges is important for their meaningful engagement in collaboratively undertaking efforts to ensure that a future for humanity on earth is not exclusively for a small number of wealthy, elite, and/or powerful humans. Taken together with insights from

CRT, conceptual tools from critical geography illuminate intersectional identity, positionality, and how one's sense of identity and positionality changes as one moves across space and time in relation to diverse others.

Postgrowth Studies

Postgrowth or degrowth studies, with roots in the work of European ecological economics scholars, engage critically with dominant, growth-oriented economic models and insist upon alternatives crucial for inclusive 'futuring' work. Some researchers who affirm degrowth or postgrowth principles, such as economist Kate Raworth, prefer other framings such as "a global economy that is regenerative by design" (Dubner, 2020, 16:42). Central tenets include recognizing the biophysical limits of the planet and the imperative of "systemic political, institutional and cultural change" for degrowth that is just and inclusive, including in intergenerational terms (Kallis, 2011, pp. 875 and 878). Postgrowth studies can, for example, offer learners of Spanish tools for understanding how the visual rhetoric of authentic materials like memes, photographs of graffiti, campaign materials, advertisements, or music videos normalizes or challenges particular cultural imaginaries around consumption, wealth, work, or the right to a particular place (the city or another locale) or mobility (travel, migration).

Environmental humanities scholarship draws upon these fields (and more) in considering how the logics and operations of power shape entangled social, spatial, and material realities in ways that negate or sustain the conditions for life, human and other-than-human, at all levels of scale and across time.

Foreign language CBI informed by critical environmental humanities theories is generative in several ways. Perhaps most apparently, it offers a critical framework through which undergraduate students can revisit and reflect upon prior learning associated with the five C's (communication, cultures, connections, comparisons, communities) of the World-Readiness Standards, which inform K-12 language education and standards (National Standards Collaborative Board, 2015). The critical framework of environmental humanities also allows graduate students in literary and cultural studies who are teaching undergraduate language classes to better connect their knowledge, interest, and expertise to their instruction and further their professional development. Finally, because environmental humanities content resonates with learning in numerous disciplines in the humanities and social sciences, as well as in pre-professional studies for health, law, and other fields, instructors may be able to envision and co-create meaningful opportunities for interdisciplinary collaboration and transdisciplinary learning, teaching, research, and public engagement. Considerations of the relevance and effectiveness of environmental humanities approaches for CBI by second language learning professionals also open possibilities for shared dialogue and deepened collaboration with colleagues in modern languages departments and professional associations, as well as multiple other fields and diverse public audiences.

Engaging 21st-Century Language Education for Shared Futures

What, though, does it look like to shape curricula, courses, or course materials in ways that acknowledge that language teaching and learning work in the 21st century happen in a pluriverse? What does it mean to have a curriculum that creates possibilities for reflection about its role in systems that can either replicate or dismantle inequities and injustices, such as institutional settings in the US? Paramount are learning experiences that foster participatory inclusion by students and critical reflection by students and instructors about positionality, intersectionality, materiality, and the uses of knowledge and power. I suggest that this work begins by acknowledging and considering working assumptions about language education; each of us can then be transparent about them to our interlocutors, including and especially our students. Among the fundamental assumptions are the following:

1   Language education in institutional settings has been shaped by the coloniality of power globally and within national cultures. In the national cultures of former colonies in the Americas, including the US, dominant cultures and dominant languages are generally settler cultures and languages that have imposed hierarchies upon all who inhabit the territory through laws, education, and imaginaries. Additionally, both media and instructional materials have often presented national cultures as linguistic monocultures when, historically, reality in diverse territories has been heterogeneous, multilingual, and pluricultural.[4]

2   Practically speaking, criticality in the curriculum begins when instructors and curriculum designers reflect on how institutional practices are rooted in the power dynamics of the cultures in which institutions are embedded and how the curricula they design and implement either maintain or work to change power dynamics for more social inclusion, greater equity, and more sustainable futures for humans and others.

3   Instructors have the opportunity to (a) imagine and design spaces in which students and instructors of diverse experiences and identities can participate with a deepening understanding of themselves in relation to diverse others and the world; (b) create courses and assignments which acknowledge the heterogeneity of lived experiences of speakers of the languages of instruction and the cultures in which that language circulates, and (c) foster metacognitive awareness of strategies and tools for teaching, learning, and engagement across difference, so that each person becomes adept at contextualizing, reflecting on, and continuously modifying their understanding of the world and their place in it. The diverse life experiences and identities of learners, their agency, and their capacity for critical reflection and growth should inform every action taken by instructors.

4   The extent of instructor autonomy is shaped by institutional contexts and constraints in both the institution of employment and professional

organizations that shape disciplinary cultures. Criticality can inform collaborative and collective work in all institutional spaces.

5   Intercultural learning involves critical reflection about learners' own identities and positionalities in relation to peers, instructors, and broader communities and members of cultures in which the language of instruction is spoken, so that cultural heterogeneity becomes more perceptible in multiple places. Language learning in a pluriverse should foster growth along a developmental pathway related to intercultural learning, from a disposition of cultural humility, and an interconnected developmental pathway related to growing competency in the language of study.

6   Finally, most importantly, language education, reimagined for just and regenerative futures, should invite students and instructors to destabilize their expectations of 'mastery' in favor of relationality and continuous learning, so that together we are predisposed to learn from others in the dynamic world in which we are enmeshed.

As instructors, if we take the first step in making perceptible, through critical reflection, frameworks that have shaped our thinking and training, we are more equipped to model and affirm skills and dispositions we encourage in our students and to create space for colleagues of identities and experiences different from our own. We are better positioned to create learning experiences that intentionally acknowledge the pluriverse, the intersectional identities of students in a pluriverse, and how the coloniality of power still operates to shape identities, knowledge, institutions, and the material world.

Sustainability-Based, Language CBI Design for the Pluriverse: Examples from Spanish

The subject matter core in curricula and courses will be emergent from numerous situational factors, and acknowledgment of situatedness is itself a step toward critical praxis. Important questions in teaching for sustainability and regeneration in the pluriverse are numerous. Who is learning? Who is teaching? Where are they learning and teaching? To what ends are people teaching and learning? What forces have historically shaped the present realities of teaching and learning within the discipline, curriculum, and institution? With clarity about situational factors and transparency about the why and how of the courses they design, curriculum designers and instructors are better prepared to teach. Transparency about the 'how' and 'why' in communication with learners is important, as it is not only effective for learning but also signals that students are invited to bring themselves into the course as co-participants in work toward learning outcomes (and in critical reflection about the outcomes themselves).

In language major courses, content is desirable that simultaneously advances the learning of linguistic, discourse, sociocultural, and action competencies (Canale, 1983; Canale & Swain, 1980; Celce-Murcia et al., 1995). Attentiveness to design is paramount so that both materials' selection and students' interaction with it further multiple outcomes. For creating and curating environmental humanities content that is grounded in both second language learning and CBI principles, it is essential to understand the intersectional identities and positionalities of the participants in learning (including the intersectional identity and positionality of the instructor), the forces that have led to their encounter with one another in the course or program, and the ways such forces are emergent from and shape enmeshed social and material worlds. Finally, it is essential, as with any course, to reflect on the desired outcomes for the course and the extent to which materials, design, and instruction further student growth toward those outcomes.

The criticality of the environmental humanities, applied to CBI design and praxis, can foster conditions for the participatory inclusion of learners of diverse identities working toward broadly embraced outcomes like multilingual proficiency, intercultural competence, and global learning. Criticality also challenges and re-situates the meaning and teleology of skills and competencies. For example, what marks a person as interculturally competent? What does it mean to be an informed global citizen? To collaborate across differences (of language and culture, among others)? To engage others equitably in creating futures in which greater thriving is possible? How does the way one answers and the way one does this work vary according to one's intersectional identity? Importantly, criticality acknowledges the entanglements of learner, social context, and materiality. In doing so, it creates the possibility for rich cognitive, affective, and relational engagement by learners and instructors with one another and with content.

Courses in the Spanish program at Wofford College make extensive use of *collaborative* learning, such as think-pair-share activities and group projects; *metacognitive* learning, for example, in oral and written reflections about one's process of learning; and the study of texts in diverse genres and modalities. The majority of students are not native or heritage speakers and enter the program at the intermediate level through the general education language requirement. Enrollments in Spanish have historically been robust at the college. Language study is popular, with as many as one in five students in each graduating class earning a language major in the early 2000s. Advanced Spanish with Community-Based Learning (CBL) occupies the curricular position of a foundational course for the major. As such, it is expected that it will offer a rich learning experience that furthers learning outcomes in speaking, reading, writing, listening, intercultural learning, and global learning. Since foundational courses are common in language curricula in the US, and this course, rather than more advanced ones in literature or culture, will be the focus of the next section.[5]

Applying Environmental Humanities to Content-Based Spanish

Advanced Spanish with CBL was re-designed as a community-engaged learning course in 2001. Community-engaged learning is a high-impact educational practice in which students learn with community partners as part of their work in a course.[6] Ideally, communities are co-creators of student learning; that is, instructors and communities are in relationship and dialogue, and instructors support meaningful student engagement with communities and foster students' critical reflection about their role in community contexts.[7] In Advanced Spanish with CBL, college students take part in regular work to support learning by primary school students in a school district comprising a high percentage of first-generation immigrant households, where a language other than English is spoken at home. As an aspirationally decolonial Spanish course grounded in environmental humanities, the course foregrounds complexity, which, as Grabe and Stoller (1997) point out, "can optimize learning in the right educational contexts" (p. 9). How, though, can the quotidian nature of content and assignments elevate complexity and further criticality?

As an overarching principle, the design of the course encourages students to become aware, or to deepen awareness, about sociospatial dynamics at different levels of scale: global, with regard to conquest, colonization, and trade; hemispheric, concerning patterns of migration; local, with regard to the presence of different racial and ethnic communities in the county and their differential access to health, safety, housing, green spaces and waterways, public services, and participatory inclusion in civic life; and personal, with regard to forces shaping their own lives and experiences. Two guiding questions for me as an instructor and lead curriculum designer are always, "How might X content and activity further work for language proficiency building and criticality in intercultural and global learning? What am I privileging, depriviling, uplifting, or centering with the inclusion of X content?" By way of illustration, the course begins with a unit about the 'why' and 'how' of the course design, with readings from higher education journals and discussion in English, so that students have tools to examine (and question) the course design as they experience it in action. This lays the groundwork for metacognitive reflection later in the course and gestures toward a decentering of instructor authority and centering of research-informed design and broader social contexts. I give here two examples of an integrative and inclusive approach to design: the first, of a unit and the second, of the sequencing and scaffolding of the curated course content.

Example 1: Design of Unit on Intersectional Identity and Positionality

The first set of Spanish language activities in the course centers around a conceptual map of intersectional identity, inclusive of categories like age; race and/or ethnicity; gender identity; relationship status; language(s); and place-oriented categories like place (state, province, country) of birth and place of citizenship or residence. Students describe their intersectional identity in writing in Spanish and give an example of how some aspect of their identity shapes their understanding of the world or how others in the world interact with and understand them. The aim is for paragraph-length discourse. Students can answer well in the present tense; those with more advanced proficiency levels can give a specific anecdote or memory that allows them to demonstrate a higher level of proficiency. After writing, students are asked to do a pair-share activity in which they talk about 3–4 aspects of identity they feel comfortable sharing and are important to them. Following that, the instructor guides students in connecting the concept of intersectional identity to positionality via scenarios or hypotheticals (i.e., does what is most present for you as an aspect of identity change for you if you are walking alone at night in [area near campus]? walking in as a guest to a meeting of college trustees? interacting with a parent or child at your community site?). This set of activities introduces key concepts, those of intersectional identity and positionality, in a way that is accessible; students' languaging suggests a growing sense of identity and an understanding that sense of identity can often be dynamic with respect to space (in particular contexts), social relationships (with diverse others), and time (across a lifetime). The set of activities also creates conditions by which there is an opportunity for all learners to understand and communicate their identities and experiences in and via the curriculum (Bianco, 2014; Lagunas, 2019). As the environmental humanities remind, there is no universal human subject, and there is no universal student, and this content early in the academic term is designed to signal that overtly.

Example 2: Design of Sequencing and Scaffolding of Content

The transparency signaled in the first unit of the course continues in the form of guides for learning through each piece of written and video content in Spanish. All videos and readings are faculty-curated, accessible for free on the internet. They have accompanying homework guides that encourage student habits for proficiency building through auditory and reading comprehension, writing and speaking, as well as attention to lexical and structural elements of language. The purpose of the guides is to make transparent processes that students can use for improving comprehension and communication within the anticipated range of levels in the class (as previously mentioned, from Intermediate-Mid to Advanced-Low). Each content unit also includes activities focused on language structures that occur in those texts and are important for Advanced level proficiency. Rounding out each unit are films, podcasts, experiential assignments, oral prompts, and critical reflection prompts for writing. Unit topics, in chronological order, are as follows: language diversity; geography and cultural history of the Iberian peninsula; geography of Latin America; conquest and colonization of Latin America; legacies of the African diaspora; colonization and nation-formation of the US; the industrial revolution; Latin American independence and nation-formation; and globalization.

The course moves from relatively familiar to less familiar content, a design that Stoller (2004), following Bragger and Rice (1999) and Bueno (2002), highlights as important for helping students further their learning without confronting proficiency jumps that are too considerable (p. 266). Moving from relatively familiar to less familiar also creates conditions by which students may connect new learning to prior learning, which is important for memory and retention. For education for greater sustainability and more inclusive socioenvironmental futures, taking the relatively familiar and reframing it so that greater complexity becomes perceptible is fundamental. Why? Simply put, many ways of being, knowing, and doing that have become normalized in education and daily life are anchored in settler-colonial logics that are exclusionary and do not acknowledge the biophysical limits of the planet.

Content related to cultural history and geography is somewhat familiar to most students. Considering it through an environmental humanities framework, however, pushes thinking beyond the superficial and toward understandings of implicit mental models, root causes, and systemic forces. In short, it encourages students to engage with the epistemological and ontological. The sequencing of the units encourages students to discern commonalities and differences throughout the hemisphere; for example, they reflect on foundational myths and visual rhetoric associated with colonialism and nation-making throughout the Americas. They also consider implications of colonialism, nation-making, and globalization for the inhabitation of territory and right to place and political participation for diverse groups of people and individuals within particular categories of identity, including themselves. Learners in the class from outside the Americas also have the opportunity to reflect upon these same topics in relation to their place of origin.

The design of both the course (unit topics, sequencing of topics, and guides) and the sequencing of activities each week invites learners to consider the logics and operations of coloniality in shaping the socioenvironmental present at all levels of scale. Each week begins with a 'big picture' topic and moves, over the week, from the highest level of scale to the more proximate and personal. For example, in the first content unit, students engage with a short expository video about language diversity in the world and the exclusion of Indigenous languages from many institutional contexts, including our own; they then read a written text about language diversity in the Americas; complete activities that focus on lexical elements and structures (present tense, connectors, numbers, and data) in both texts; and conclude the week with content about language heterogeneity in their city and their lives. Critical reflection with guiding prompts encourages students to deepen their understanding of themselves and others as persons of diverse identities, whose lives are shaped by social forces, over time, in the multiple spaces where they live and learn.

In the interest of furthering efforts for equity and consideration of local topics, the course includes English-language content about local contexts (because Spanish language content on local topics is often simply unavailable) and offers students the possibility of written critical reflection in English about community-engaged learning (or multimedia alternatives for those for whom

CBL is not possible at a given time, for example, because of illness). Creating space for critical reflection in English signals to students that they may convey their deepest learning (including application and metacognitive learning) without stretching their proficiency beyond the point of language breakdown. It also furthers students' translingual and transcultural competence, defined by a 2007 Modern Language Association (MLA) report as an "ability to move between languages" and to learn and reflect upon "differences in meaning, mentality, and worldview as expressed in American English and the target language" (Geisler et al., 2007, p. 237). The MLA report presents translingual and transcultural competence goals as a paradigm shift for language instruction and a necessary movement away from an unrealistic goal of instruction to replicate the competence of a native speaker.

Assessment of learning gains is through labor-based, contract grading,[8] which emphasizes methodical efforts at learning and improving proficiency through application of the guides and allows the instructor to be responsive to the ranges of proficiency among students in the course. Students receive the instructor's feedback on work completed weekly via a rubric they receive at the beginning of the course. Though always a work in progress, the course design seeks to foster a deeper, relational engagement with course material and disincentivize an extractive, instrumentalist approach to coursework.

Taken as a whole, design, content, and activities all ask students to engage in scalar thinking (Bennett et al., 2007) in relation to the *longue durée* of history. Dynamics and differentials of power vis-à-vis human, material, and spatial relationships become foregrounded in ways that encourage students to think deeply about the constructedness of the present and the quotidian.[9] The community-engaged component, in particular, encourages students to perceive and articulate how global forces act in local spaces and re-shape neighborhoods, urban life, and public spaces.

Teaching in such a way is not easy: it requires holding present in mind multiple outcomes at once and designing materials that further growth in each. It also demands a level of comfort with student agency and autonomy and some training and expertise in handling discussions in the classroom that produce discomfort for some students. Finally, a course inclusive of local levels of scale and community-engaged learning is predicated upon being-in-relationship with communities and deep and frequent listening. Doing all of this at once also means continuous reflection upon one's assumptions: about language, discipline, and profession; about institutional and social roles; about dynamics of power and what it means to navigate power for inclusive human futures on the planet.

Conclusion: Toward Sustainability in the Pluriverse

As urban theorist Edward Soja (2010) writes, "[e]verything is connected to everything else, but not just in a flat horizontal ecosystem or biosphere. These connections also extend vertically through a socially produced layering of bounded geographical scales extending from the planet to the body" (p. 54).

A modern languages curriculum is well-suited to undertake the daily, iterative, collaborative work to direct students' attention to multiple scales (from the body to the global scale) in which interventions for more just and sustainable futures may occur. Modern languages curricula informed by the environmental humanities have the potential to engage a globally mobile learner understood in terms of positionality, intersectionality, materiality, and spatiality, as a person who exerts agency and upon whom diverse forces are exerted in multiple places and ways and to different degrees. Such curricula hold space for the complexity of lived reality for students and instructors and acknowledge the intersectional forces that shape our individual lives and shared social futures. Through engagement with the environmental humanities, the many purposes of language education can begin to be articulated, not just by instructors and curriculum designers, but by students themselves in critical reflection, dialogue, and action with others. The process of re-imagination, ideation, co-creation, and iterative implementation of curricula in a context of heterogeneity and with an understanding of scale, moves us away from educational monocultures (Taylor, 2008) and into a pluriverse in which more just, regenerative futures are possible. Such teaching and learning awaken us to our roles as co-creators of a world in the making.

Notes

1 See, for example, the AAC&U (2009) Intercultural Competence and Knowledge rubric, which references "history, values, politics, communication styles, economy, or beliefs and practices" as aspects of cultural difference and does not explicitly mention biophysical realities that are shaped by (and shape) cultures.
2 Many other fields inform critical work in the environmental humanities, including postcolonial studies, science and technology studies, posthumanism, animal studies.
3 In Latin American contexts, decolonial theory has been shaped by the scholarship of Walter Mignolo, María Lugones, Aníbal Quijano, Enrique Dussel, Ramón Grosfoguel, Arturo Escobar, Marisol de la Cadena, and Santiago Gómez-Castro, among others.
4 For a profoundly poetic engagement of the "colonization of the imagination" by national borders, see Yásnaya Elena Aguilar Gil's (2020) "The Map and the Territory" (https://thebaffler.com/latest/the-map-and-the-territory-aguilar-gil)
5 In our curriculum, approximately half of the students enrolled in the sections of Advanced Spanish with CBL are first-year students, and half are second-year students.
6 For more about civic engagement, including community-based and community-engaged learning, see Saltmarsh, J., & Zlotkowski, E. (2011). *Higher education and democracy: Essays on service-learning and civic engagement.* Temple University Press.
7 For more about the creation of Advanced Spanish with Community-Based Learning, see Barbas Rhoden, L., & Tate, J. (2014). Connecting Curriculum to Context: Our Story of Two Liberal Arts College Spanish Programs Engaged in a Changing South. *PRISM: A Journal of I Engagement, 3* (1). https://encompass.eku.edu/prism/vol3/iss1/4.
8 For more about labor-based grading and other forms of ungrading, see Blum, S. D., & Kohn, A. (2020). *Ungrading: Why rating students undermines learning (And what to do instead).* West Virginia University Press.

9 The choice to further relational engagement is informed by critical environmental humanities theory and research about second language learning. Kennedy (2006), in discussing the implications for language instruction of studies on brain chemistry and structures, points out that "enriched environments promote neuronal development," emotions "drive attention" and tell us "what is important to learn and what to remember," and that the brain "stores information based on functionality and meaningfulness" (p. 479). She asserts that curriculum designers must provide learners with "higher-level activities promoting application, analysis, synthesis, and evaluation" (p. 479). Her insights help explain why courses with experiential components advance learning and why relational processes for integrative learning are important.

References

Aguilar Gil, Y. E. (2020). *The map and the territory*. The Baffer, New York University. https://thebaffler.com/latest/the-map-and-the-territory-aguilar-gil

Association of American Colleges and Universities (AAC&U). (2009). *Intercultural knowledge and competence value rubric*. https://www.aacu.org/value/rubrics/inquiry-analysis

Barbas Rhoden, L., & Tate, J. (2014). Connecting curriculum to context: Our story of two liberal arts college Spanish programs engaged in a changing south. *PRISM: A Journal of Regional Engagement, 3*(1), 57–66. https://encompass.eku.edu/prism/vol3/iss1/4

Bennett, D. C., Cornwall, G. H., Al-Lail, H. J., & Schenck, C. (2017). *An education for the twenty-first century: Stewardship of the global commons*. Association of American Colleges and Universities. https://www.aacu.org/publications-research/periodicals/education-twenty-first-century-stewardship-global-commons

Bianco, J. (2014). Domesticating the foreign: Globalization's effects on the place/s of languages. *The Modern Language Journal, 98*(1), 312–325. https://doi.org/10.1111/j.1540-4781.2014.12063.x

Blum, S. D., & Kohn, A. (2020). *Ungrading: Why rating students undermines learning (and what to do instead)*. Morganton, WV: West Virginia University Press.

Bragger, J. D., & Rice, D. B. (1999). The message is the medium: A new paradigm for content-oriented instruction. *Foreign Language Annals, 32*(3), 373–391. https://doi.org/10.1111/j.1944-9720.1999.tb01349.x

Bueno, K. A. (2002). Creating community and making connections in the third-year Spanish course: A content-based approach. *Foreign Language Annals, 35*(3), 333–342. https://doi.org/10.1111/j.1944-9720.2002.tb01858.x

Canale, M. (1983). From communicative competence to communicative language pedagogy. *Language and Communication, 1*(1), 1–47.

Canale, M., & Swain, M. (1980). Theoretical bases of communicative approaches to second language teaching and testing. *Applied Linguistics, 1*(1), 1–47. https://doi.org/10.1093/applin/I.1.1

Celce-Murcia, M., Dörnyei, Z., & Thurrell, S. (1995). Communicative competence: A pedagogically motivated model with content specifications. *Issues in Applied Linguistics, 6*(2), 5–35.

Crenshaw, K. (1990). Mapping the margins: Intersectionality, identity politics, and violence against women of color. *Stanford Law Review, 43*(6), 1241–1299. https://doi.org/10.2307/1229039

Davis, H., & Todd, Z. (2017). On the importance of a date, or decolonizing the Anthropocene. *ACME: An International Journal for Critical Geographies, 16*(4), 761–780.

Delgado, R., & Stefancic, J. (2001). *Critical race theory: An introduction.* New York: NYU Press.

Dubner, S. (Host). (2020, August 12). Is economic growth the wrong goal? Ep. 429 [Audio podcast episode] In *Freakonomics.* Freakonomics Radio. https://freakonomics.com/podcast/doughnut-economics/.

Escobar, A. (2018). *Designs for the pluriverse.* Durham, NC: Duke University Press.

Geisler, M., Kramsch, C., McGinnis, S., Patrikis, P., Pratt, M., Ryding, K., & Saussy, H. (2007). Foreign languages and higher education: New structures for a changed world: MLA ad hoc committee on foreign languages. *Profession,* 234–245. http://www.jstor.org/stable/25595871

Gómez-Barris, M. (2017). *The extractive zone.* Durham, NC: Duke University Press.

Grabe, W., & Stoller, F. L. (1997). Content-based instruction: Research foundations. In D. Brinton & M. A. Snow (Eds.), *The content-based classroom: Perspectives on integrating language and content* (pp. 5–21). Ann Arbor, MI: University of Michigan Press.

Jazeel, T. (2017). Mainstreaming geography's decolonial imperative. *Transactions of the Institute of British Geographers, 42*(3), 334–337. https://doi.org/10.1111/tran.12200

Kallis, G. (2011). In defense of degrowth. *Ecological Economics, 70*(5), 873–880. https://doi.org/10.1016/j.ecolecon.2010.12.007

Kennedy, T. J. (2006). Language learning and its impact on the brain: Connecting language learning with the mind through content-based instruction. *Foreign Language Annals, 39*(3), 471–486. https://doi.org/10.1111/j.1944-9720.2006.tb02900.x

Lagunas, R. M. (2019). Nahuatl in Coatepec: Ideologies, practices and management for linguistic and cultural continuance. *International Review of Education / Internationale Zeitschrift Für Erziehungswissenschaft, 65*(1), 67–86. https://doi.org/10.1007/s11159-019-09765-6

National Standards Collaborative Board. (2015). *World-readiness standards for learning languages* (4th ed.). National Standards Collaborative Board.

O'Gorman, E., Van Dooren, T., Münster, U., Adamson, J., Mauch, C., Sörlin, S., Armiero, M., Lindström, K., Houston, D., Augusto Pádua, J., Rigby, K., Jones, O., Motion, J., Muecke, S., Chang, C., Lu, S., Jones, C., Green, L., Matose, F.,… Jørgensen, D. (2019). Teaching the environmental humanities: International perspectives and practices. *Environmental Humanities, 11*(2), 427–460. https://doi.org/10.1215/22011919-7754545

Prádanos, L. (2018). *Postgrowth imaginaries: New ecologies and counterhegemonic culture in post-2008 Spain.* Liverpool: Liverpool University Press.

Price, P. L. (2010). At the crossroads: Critical race theory and critical geographies of race. *Progress in Human Geography, 34*(2), 147–174. https://doi.org/10.1177%2F0309132509339005

Quijano, A. (2000). Coloniality of power and eurocentrism in Latin America. *International Sociology, 15*(2), 215–232. https://doi.org/10.1177%2F026858090000 15002005

Rodgers, D. M. (2014). Making the case for content-based instruction. *Italica, 91*(1), 16–28.

Saltmarsh, J., & Zlotkowski, E. (2011). *Higher education and democracy: Essays on service-learning and civic engagement.* Philadelphia, PA: Temple University Press.

Sato, S., Hasegawa, A., Kumagai, Y., & Kamiyoshi, U. (2017). Content-based instruction (CBI) for the social future: A recommendation for critical content-based language instruction (CCBI). *L2 Journal, 9*(3), 50–69. https://doi.org/10.5070/L29334164

Soja, E. W. (2010). *Seeking spatial justice*. Minneapolis, MN: University of Minnesota Press.

Stoller, F. L. (2004). Content-based instruction: Perspectives on curriculum planning. *Annual Review of Applied Linguistics, 24*, 261–283. doi:10.1017/S0267190504000108

Stryker, S. B., & Leaver, B. L. (1997). *Content-based instruction in foreign language education: Models and methods*. Washington, D.C.: Georgetown University Press.

Taylor, L. K. (2008). Of mother tongues and other tongues: The stakes of linguistically inclusive pedagogy in minority contexts. *Canadian Modern Language Review, 65*(1), 89–123. https://doi.org/10.3138/cmlr.65.1.89

Tuck, E., & Yang, K. W. (2018). Introduction by the journal coeditors. *Critical Ethnic Studies, 4*(1), 1–3. doi:10.5749/jcritethnstud.4.1.0001

Whyte, K. (2017). Indigenous climate change studies: Indigenizing futures, decolonizing the Anthropocene. *English Language Notes, 55*(1), 153–162. https://doi.org/10.1215/00138282-55.1-2.153

2 Multiliteracies Pedagogy

Theory to Practice for Scaffolding Sustainability Literacies

Kristen M. Turpin

Introduction

Content-based instruction (CBI) is a curricular approach that embeds language instruction in the context of meaningful content for learners (Brinton & Snow, 2017; Cammarata, 2016a; Met, 1999; Snow et al., 1989; Stoller, 2004). Because it offers a framework for creating cognitively engaging instructional materials that develop learners' linguistic proficiency (Genesee & Lindholm-Leary, 2013), CBI is transforming L2 classrooms into a space where students use language to inquire about global issues that are most pressing to them, such as climate change. Fortunately, sustainability is a topic that lends itself to CBI and can be integrated at various levels. Sustainability studies can be connected to students' daily lives yet demand that they engage in higher-order thinking, consider cultural differences, and make connections to other disciplines (Kautz, 2016). In terms of plausibility, teaching about sustainability in the L2 content-based classroom also has its advantages. CBI depends on the use of foreign language texts "to promote purposeful language use, understanding of complex meaning, and extended learning activities as beneficial to learner discovery of the multiple dimensions of language and culture" (Melin, 2019, p. 303). Because the internet abounds with infographics, short videos, articles, and government-sponsored information campaigns directed at populations with varying degrees of literacy, instructors can find relevant, accessible texts to provide a meaningful context in which students communicate purposefully about global sustainability issues.

Beyond providing the content and context in a sustainability-based course, multimodal texts of various genres play a central role in redesigning college-level foreign language education to meet the needs of 21st-century learners. As departments continue to integrate the study of language, literature, and culture at all levels of the curriculum, educators and researchers alike are reconsidering the types of texts that L2 learners should encounter and how, when, and why they should interact with those texts. The *multiliteracies framework* is one of the leading approaches because it integrates "the teaching of 'communication' with the teaching of 'textual analysis'" (Kern, 2003, p. 43) in order to prepare foreign language learners "to participate in diverse discourse communities both at home and in the target culture" (Paesani et al., 2016, p. 3). The multiliteracies

DOI: 10.4324/9781003080183-4

framework (also known as 'multiliteracies pedagogy') emphasizes the socially- and culturally-situated nature of language and communication, inviting learners to analyze how texts are produced in their contexts of use (Kern, 2000). Consequently, multiliteracies pedagogy asks learners to make "connections between grammar, discourse, and meaning" (Kern, 2000, p. 46), which transforms instruction in two ways. First, learning objectives shift from the comprehension of key concepts to analyzing how and why different cultures, disciplines, news outlets, and stakeholders construct texts about humans' relationship with the natural world. Second, grammar is no longer a list of rules to memorize but "a system of discourse tools" (Swaffar & Arens, 2005, p. 21) that writers/speakers use to describe, debate, and inquire about climate change and environmental justice.

After introducing the multiliteracies framework and justifying its relevance to sustainability-based instruction, this chapter illustrates how to combine CBI and multiliteracies pedagogy in a fifth-semester Spanish conversation course. Drawing examples from a unit on 'Ecotourism and Environmental Conservation,' I demonstrate how to reframe reading and writing as socially- and culturally-situated acts of meaning design. I explain how to move learners through the four knowledge processes that constitute literacy-based learning—experiencing, conceptualizing, analyzing, and applying (Kalantzis et al., 2016, p. 67)—as they explore competing and contradicting narratives of ecotourism in Costa Rica.

Multiliteracies Pedagogy: Developing Sustainability Literacies

It is well documented that CBI has a variety of faces. It encompasses a wide range of curricular models, from content-driven programs in which "student learning of content is of greater importance than language learning" to language-driven programs in which "content is a useful tool for furthering the aims of the language curriculum" (Met, 1999, para. 6). Although institutional and departmental structures often determine where content-based foreign language courses lie on Met's spectrum of language-content integration, it is a valuable exercise to think hypothetically about the consequences of teaching about sustainability at various points along his spectrum. At one extreme, students in content-driven courses would gain information about how human activity affects the environment, but they might struggle to put their knowledge into practice and do something—with language—to enact change. In other words, they might lack sustainability literacy, or "the collection of skills" that allow individuals to participate effectively in efforts to build a "more sustainable self, community, society, and world" (Stibbe & Luna, 2009, p. 1). At the language-driven end of Met's spectrum, "theme-based courses" and "language classes with frequent use of content for language practice" might also fall short of empowering students to take climate action. In a language-oriented classroom, students might not be held accountable for content mastery (Met, 1999, para. 10). Furthermore, novice or intermediate learners might not be challenged to stretch their language proficiency beyond communicating about the self. Although the personal choices that we make to reduce our carbon footprint do impact the future of our planet, teaching and learning about sustainability in the L2 classroom should not stop there. In its most radical sense, sustainability literacy is collaborative and revolutionary; in addition to reducing our individual carbon footprints, we need to take on the systems of power that prevent real progress toward more sustainable futures.

In the middle of Met's spectrum, there are CBI courses that equally prioritize content and language learning, and these often succeed in preparing students to participate in academic discourse communities (Met, 1999, para. 15). However, solving the climate crisis and pursuing climate justice requires buy-in from various constituent groups, both inside and outside of academia. Students need the ability to translate their content knowledge, present it persuasively in various contexts, and build relationships among groups with conflicting perspectives and competing interests. Especially in the context of sustainability, L2 learners need to develop a purpose-sensitive and agentive sense of multiple literacies—which is what

> empowers individuals to enter societies; to derive, generate, communicate, and validate knowledge and experience; to exercise expressive capacities to engage others in shared cognitive, social, and moral projects; and to exercise such agency with an identity that is recognized by others in the community.
>
> (Swaffar & Arens, 2005, p. 2)

Content-based courses that prioritize academic literacy might fall short of empowering students in this multifaceted way. Therefore, this chapter argues that sustainability-based L2 education, regardless of the relative value assigned to the learning of language and content, should consider the different types of literacies that one may possess and the different contexts in which literacies practices take place (Paesani et al., 2016, pp. 11–12; The New London Group, 1996). To emphasize this key point, sustainability literacies will be referenced in the plural.

Given the conceptual compatibility of 'sustainability literacy' and 'multiple literacies,' L2 courses that aim to develop students' sustainability literacies often complement CBI with models for multiple literacies teaching. One of these models is the multiliteracies framework, which embraces the plural, multifaceted definition of literacy mentioned in the previous paragraph and aims to reflect the "dynamic, socially-determined, and multidimensional uses of language encountered in daily life" (Center for Advanced Research on Language Acquisition, 2021). The origins of the multiliteracies framework can be traced to debates in the 1980s and 1990s about the limits of communicative language teaching (CLT), understood here as an instructional approach that focuses on interactive, transactional oral language usage in generic contexts

(Allen & Paesani, 2010, p. 122). One of the main critiques of CLT is the limited role of textual content. Viewed as secondary support skills, reading serves to deliver comprehensible input, and writing serves to practice language forms; neither modality offers the opportunity to interpret cultural content, create written texts with real social purposes, or develop thinking and intellectual abilities (Allen & Paesani, 2010, p. 122; Paesani et al., 2016, p. 7; Warner & Dupuy, 2018, p. 117). These critiques laid the groundwork for the New London Group's seminal publication, which proposed a 'pedagogy of multiliteracies' that would enable students to negotiate linguistic and cultural differences in increasingly globalized societies (The New London Group, 1996).

Following the New London Group's redefinition of language learners as "active designers of meaning" (p. 65), Kern (2000) provided practical models for teaching reading and writing as interconnected, dynamic acts of meaning design. When the Modern Language Association (MLA) punctuated the need to teach "language, culture, and literature as a continuous whole" (Modern Language Association, 2007), the multiliteracies framework offered a concrete method for implementing program-wide curricular reform. Swaffar and Arens (2005) implemented the multiliteracies framework to integrate the study of literature and culture into every level of the curriculum and redefine the role of foreign language programs within the humanities. Allen and Paesani (2010) explored its feasibility in introductory foreign language courses, and Byrnes et al. (2010) used genre-based literacies instruction to develop students' advanced writing proficiency. Paesani et al. (2016) synthesized decades of scholarship in an accessible theory-to-practice guide for educators. Recently, Paesani (2018) and Warner and Dupuy (2018) have looked toward the future of multiliteracies pedagogy and identified new context and focal points, such as heritage language education and social justice education. By highlighting the relevance of multiliteracies pedagogy to sustainability-based L2 education, this chapter illustrates one such possibility for the future of multiliteracies.

Some sustainability-minded curriculum designers intuitively combine aspects of CBI and multiliteracies (de la Fuente, 2019; ter Horst & Pearce, 2010), whereas others explicitly and intentionally adopt practices from the multiliteracies framework (Kautz, 2016; Melin, 2013, 2019). Méndez Seijas and Luisa Parra (this volume) adapted select aspects of multiliteracies pedagogy to make complex sustainability content accessible for novice learners. Working at the intermediate level, Kautz (2016) supplemented a textbook with sustainability-themed CBI lessons to develop students' multiple literacies—"the skills and attitudes that make learners able to participate fully in the life of different communities" (Kautz, 2016, p. 263). When Melin (2013) designed an upper-division course on Contemporary Germany, she recognized that "learning about environmental content would be just as important to the educational process as textual interpretation," so she intentionally blended "models for multiple literacies teaching" and CBI (Melin, 2013, p. 187). In *Foreign language teaching and the environment* (2019), Melin advocated "attention to multifaceted work on literacy instead of concentrating on speaking skills" when teaching sustainability and environmental humanities topics in language, literature, and culture courses (Melin, 2019, p. 5).

The following pages illustrate a theory-to-practice approach to using a multiliteracies framework to scaffold students' development of sustainability literacies. This approach builds upon existing scholarship by rendering transparent key decisions that must be made in the process of multiliteracies curriculum development. In order to exemplify this decision-making process, I reference a unit about ecotourism in Costa Rica that was developed for a special topics section of a fifth-semester Spanish conversation course at Brandeis University. The course, which has the overarching goal of developing oral proficiency at the advanced-low level, is a "sustained content course" because students explored the theme of sustainability throughout the semester (Brinton & Snow, 2017, p. 9).

Unit design: Instructional Approaches, Learning Objectives, L2 Texts

In developing this 15-hour (approx.) instructional unit, backward design was used to determine learning objectives, design assessment tasks, and plan text-based instructional activities (Wiggins & McTighe, 2005).[1] Strategies from CBI were implemented to determine the specific knowledge needed to be introduced in Phase 1, write content objectives that specified the depth of cognitive engagement, and identify the language needed for students to manipulate the targeted concepts (Cammarata, 2016b). In Phases 2 and 3, the multiliteracies framework helped strategize the day-to-day planning of instructional activities that moved students from literal comprehension to complex text-based thinking. Table 2.1 summarizes the learning objectives, L2 texts, and tasks of the unit. While the main objectives of each phase were directly assessed via formal task-based assessments and a pre-/post-unit survey, subsidiary objectives were assessed indirectly and informally via instructional activities.

Throughout the unit, students engage with foreign language texts that serve different purposes in developing students' sustainability literacies. Students first encounter short, informational texts from the sciences and social sciences. The podcast, infographic, and websites in Phase 1 allow students to define ecotourism and then explain why Costa Rica is a popular destination. To be successful, sustainability projects need to be rooted in the local—the specific ecological, economic, and sociocultural factors that shape what is considered sustainable for a given community—and the texts in Phase 1 start to provide this local specificity. Students were not asked to produce their own podcasts, infographics, or websites, so these texts were not models of language usage but rather a means of introducing concepts. Given this focus on content, basic scaffolding techniques from CBI, such as activating prior knowledge and using graphic organizers, were sufficient to support student learning at the advanced-low level (Cammarata, 2016a).

Table 2.1 Overview of unit: ecotourism and environmental conservation

Phase	Learning objectives	Instructional texts	Tasks: activities and assessments
1: Introduction — What is ecotourism? Why is Costa Rica such a popular destination?	**Main**: Explain why C. Rica is a popular destination for ecotourism. **Subsidiary:** • Define ecotourism; compare it to traditional tourism. • Compare tourists' options when vacationing in C. Rica; express your preferences and make compromises with your travel companions.	• Podcast about ecotourism. • Infographic on ecotourism in C. Rica. • Websites: traveling to the Caribbean Coast of C. Rica.	• Formal interpersonal speaking: Work with a peer to plan a trip to the Caribbean Coast of C. Rica. Students record an improvised 10-minute conversation, and their performance is evaluated with an analytic rubric.
2: Deep Dive — How is ecotourism implemented in C. Rica? Does it have any problems?	**Main**: Describe the specific challenges of implementing ecotourism in Costa Rica. **Subsidiary:** • Identify arguments for and against 'sustainably' developing the wildlife refuge Gandoca-Manzanillo. • Analyze interests and motives of different stakeholders in the debate about Gandoca-Manzanillo.	• Infographic from World Wetlands Day. • News video about the economic importance of wetlands in C. Rica. • Brochure for tourists visiting Gandoca-Manzanillo. • *La loca de Gandoca*, a novel by A. Rossi. • Interview with Rossi.	• Short, informal interpretive reading tasks related to *La loca de Gandoca*. • Formal interpretive listening: Listen to an interview with Rossi to inform your reading of the novel /hear another perspective about the challenges of implementing ecotourism.
3: Application	**Main**: Argue different points of view in debates about ecotourism and environmental conservation. **Subsidiary:** • Persuade others to share your opinion. • Reflect on your role as a tourist.	• Model Spanish-language petitions from Change.org.	• Formal presentational writing: Assume the role of one of the characters in *La loca de Gandoca*; write a petition to argue for/against the development of the wildlife refuge. The task is formally graded (analytic rubric). • In-class interpersonal speaking activity: Revise the trip you 'planned' to C. Rica to account for what you have learned about ecotourism. What would you have done differently?

Beyond informational texts that present a single perspective on key concepts, it is essential that multiliteracies curricula also incorporate texts that invite students to analyze how language is used to construct a plurality of often contradicting perspectives on the same issue (Byrnes et al., 2010, p. 123). If students are to develop dynamic sustainability literacies that empower them to participate in a variety of discourse communities, they need to examine how meaning "is constructed through form in a cultural context" (Kern, 2000, p. 185). In this unit, a longer, literary text provided this crucial opportunity for textual analysis anchored in a sociocultural context.

Anacristina Rossi[2] published the testimonial novel *La loca de Gandoca (The Madwoman of Gandoca)* in 1992 to recount her fight against the 'sustainable' development of the Gandoca-Manzanillo Wildlife Refuge, which encompasses lowland rainforests, wetlands, and mangrove swamps on the Caribbean Coast of Costa Rica (Rossi, 2009). The novel aligned well with the unit's learning objectives because it allowed students to analyze how residents, conservationists, government officials, and foreign investors crafted different arguments about sustainable development in the name of ecotourism. Instead of formally assessing students' comprehension as they were reading, multiliteracies pedagogy emphasized active, collaborative meaning-making processes. Engaging with the perspectives presented in *La loca de Gandoca* prepared students for the presentational writing task that concluded the unit: Students assumed the role of one of the characters in the novel and wrote a petition for/against the development of tourist infrastructure in the wildlife refuge.

Consequently, the learning objectives for Phase 2 require students to engage in more complex text-based thinking that recognizes the social, historical, and cultural dimensions of knowledge production and communication. In other words, students not only identify the arguments for and against 'sustainably' developing Gandoca-Manzanillo (the 'what') but also analyze how each argument gains or loses traction with community members (the 'how' and the 'why'). The multiliteracies framework's emphasis on context-sensitive, purpose-driven communication provides the theoretical coordinates and concrete instructional strategies necessary to scaffold the development of sustainability literacies in Phases 2 and 3 of the unit. The following two sections examine the pedagogical impact of two key instructional strategies: Defining reading as an act of meaning design and intentionally guiding students to engage in various knowledge processes as they interact with L2 texts.

Multiliteracies Pedagogy: Reading about Sustainability as an Act of Meaning Design

Within the multiliteracies framework, literacy extends beyond the static ability to read and write and encompasses the "dynamic, culturally and historically situated practices of using and interpreting diverse written and spoken texts to fulfill particular social purposes" (Kern, 2000, p. 6). Building upon Swaffar and Arens' (2005) empowering notion of multiple literacies, which was quoted earlier in this chapter, the multiliteracies framework uses the term 'meaning design' to emphasize how language users exercise their agency as they interact with texts and participate in society (Cope & Kalantzis, 2009; Paesani et al., 2016; The New London Group, 1996). In the context of this unit, students designed meaning as both readers and writers. In Phase 2, they created meaning through an interactive interpretation of *La loca de Gandoca*, and then, in Phase 3, they used the knowledge that they gained through textual interaction to design meaning as recognized members of various discourse communities.

Reframing the reading of this novel as an act of meaning design impacted instruction in two ways; first, instructional materials needed to consider the cognitive, linguistic, and sociocultural dimensions of reading, and second, students needed to engage in reading processes that honored the interactive nature of constructing meaning from a text (Paesani et al., 2016, pp. 143–144). To address the cognitive dimension of reading, students received explicit instruction on processes, skills, and problem-solving techniques that would facilitate a more productive interaction with the novel (Kern, 2000). For example, one of the cognitive challenges presented by the novel is that it recounts the attempt to develop the wildlife refuge through dialogue as opposed to narration. To address this, class time was dedicated to having students read the dialogue aloud, discussing the textual clues that helped them identify the speaker of each utterance, and then writing characters' names in the margins. Although the novel poses linguistic challenges such as precise legal terminology and references to local plants and animals, instructors can prioritize the meaning-making tools that best align with the communicative needs of their students. For example, the reading guides for each section of the novel focused students' attention on the lexical features used to construct arguments about ecotourism; these were the same lexical features that students would use to design meaning in the presentational writing task.

However, within the multiliteracies framework, reading extends beyond its cognitive and linguistic dimensions; "it is a set of patterned literacy practices that are tied to and reflect a group, community, or culture" (Paesani et al., 2016, p. 144). Instructional materials and in-class activities addressed the sociocultural dimension of reading by analyzing the choices that Anacristina Rossi makes in *La loca de Gandoca*. For example, the novel interweaves two parallel stories: A personal story, in which the author builds upon the literacy practices of poets to recount the destruction of her marriage, and a public story, in which the literacy practices of lawyers and public officials are transformed to recount the destruction of the wildlife refuge Gandoca-Manzanillo (Kearns, 1998, p. 315). Although students were able to identify the differences in content, style, and purpose of the personal and private stories, they struggled to correctly contextualize them within the literacy practices of a group, community, or culture. For example, students knew that the author/protagonist was a local resident, and that some local residents were English speakers of Afro-Caribbean descent; as a result, some students incorrectly assumed that Anacristina Rossi was Black and that she wrote *La loca de Gandoca* as a member of the Afro-Caribbean community. This was a clear example of how meaning-making—both on the part of the text and the part of the reader—is contextually bound; texts and the way they are (mis)interpreted reflect and propagate cultural perspectives (Kern, 2008, p. 367; Rowland et al., 2014, p. 140). Listening to an interview with Anacristina Rossi (as an interpretive listening assessment task) helped students understand the communities with which the author identifies and then situate the literacy practices of the novel accordingly. In sum, by considering the cognitive, linguistic, and sociocultural dimensions of reading, instructors can design instructional materials that support students as they overcome the challenges that a text poses for meaning design.

The multiliteracies framework provides additional strategies for teaching challenging texts by emphasizing the interactive nature of constructing meaning with texts. Within the multiliteracies view of reading, readers actively negotiate with a text's meaning by interacting with it: applying one's background knowledge, making predictions, discussing the text with peers, doing additional research, and revisiting the text multiple times, all contribute to meaning construction (Paesani et al., 2016, pp. 141–142). In order to encourage active, interactive reading practices, the novel was divided into five sections, and focus questions were designated for each section; these questions centered students' attention on the passages that would help them achieve the unit's learning objectives (see Table 2.1). The class was organized into reading groups, each responsible for answering one of the focus questions and bringing textual evidence to class. During class time, each group had 15–20 minutes to refine their understanding of the focus question and prepare a summary for their peers. For the next 20–25 minutes, each group shared their summary with the class, cited textual evidence, and answered any follow-up questions. As the instructor, I intentionally gave the class the chance to co-construct meaning, and I only intervened to provide background knowledge that students could not deduce from textual analysis and needed to advance their discussion. After all groups presented, I debriefed their contributions by pointing students to key passages that would correct any misinterpretations and summarizing how their understanding of ecotourism and environmental conservation was evolving.

To review, the multiliteracies framework considers reading to be an act of meaning design and thus provides a framework for creating instructional materials that address the cognitive, linguistic, and sociocultural dimensions of active, interactive reading practices. Consequently, it supports L2 readers throughout the interpretation of longer, more complex texts that broach various perspectives in sustainability debates. The following section addresses the remaining important questions: How does the understanding of texts translate to the real-world ability to collaboratively build more sustainable futures? When students interact with teacher-generated literacy materials, how can we be sure they are engaging in the complex text-based thinking that develops sustainability literacies?

Multiliteracies Pedagogy: Meaning-Making Actions That Build Sustainability Literacies

In multiliteracies pedagogy, the Knowledge Process (KP) framework describes the concrete types of "thinking-in-action" that students do to design meaning as they interact with texts (Kalantzis et al., 2016, p. 74). This framework can be used as a tool for literacies-oriented course planning and materials design or to analyze existing instructional materials (Fterniati, 2010; Menke & Paesani, 2019; Rowland et al., 2014). According to this framework, the four "epistemic moves, or things students can do to know" as they interpret and create texts are experiencing, conceptualizing, analyzing, and applying (Kalantzis et al., 2016, p. 80). Experiencing involves "spontaneous, immersive learning without conscious reflection"; conceptualizing consists of students actively "unpack[ing] the language forms, conventions, organizational features, and form-meaning relationships that characterize texts"; analyzing entails "relating textual meaning to social, cultural, historical, or ideological contexts and purposes"; and applying occurs when students "use new understandings and skills and produce language in conventional or creative ways" (Menke & Paesani, 2018, p. 4). Although these four knowledge processes constitute different meaning-making actions, they are neither hierarchal nor sequential, and they can be implemented in whatever order best meets students' needs (Kalantzis et al., 2016; Menke & Paesani, 2019).

For this project, I used this framework to design text-based instructional activities that would engage students in the meaning-making actions that would promote the development of sustainability literacies. Table 2.2 shows how students circled through various stages of literacies-oriented reading instruction, both inside and outside the classroom and across multiple class sessions (see Paesani et al., 2016). Most notably, students engaged in all four knowledge processes before, during, and after reading *La loca de Gandoca*. In contrast to communicative language teaching, in which experiencing predominates, literacies-oriented instructional materials should include an appropriate mix of knowledge processes to maximize proficiency gains, critical thinking, and student learning (Menke & Paesani, 2018).

The frequency of analyzing activities is fundamental for developing students' sustainability literacies. In the two critical reading stages of the unit (after reading the first half of the novel, and then after finishing it), students analyze the sociocultural context in which beliefs about sustainability are produced and cemented. For example, the analyzing activity that students completed after reading Parts 1 and 2 of the novel asked them to connect each character's views about ecotourism to their interests, experiences, and beliefs. Students identified competing ecological, economic, and sociocultural concerns, all of which must be addressed in truly sustainable projects. Additionally, they found numerous examples of racist conceptualizations of 'development' and 'progress,' as well as sexist power dynamics that determined whose views about ecotourism were legitimized and whose were not. Although

Table 2.2 Stages of reading instruction and knowledge processes in Phases 2 and 3

	What students do	*Knowledge processes*
Pre-reading	Read an informational pamphlet about the wildlife refuge Gandoca-Manzanillo; study an infographic about the ecological importance of wetlands; watch a news report about the environmental importance of wetlands. Make a concept map that answers the question: What is the value of Gandoca-Manzanillo? Why is it important?	Experiencing and analyzing
	View the cover of the novel; make predictions about what the novel's plot.	Experiencing and analyzing[3]
	Preview a passage from the novel; identify textual clues that indicate whether this type of ecotourism exists in the region.	Experiencing and conceptualizing
	Before reading the first part of the novel, read a list of key events and sequence them in logical order.	Experiencing
Initial reading	Parts 1 and 2: Read with a focus on the comprehension question assigned to the group.	Experiencing
Detailed reading	Parts 1 and 2: Revisit key passages; present the assigned question.	Experiencing
Critical reading	With selections of Parts 1 and 2: Complete a collaborative text annotation activity to connect each character's views about ecotourism to their interests and experiences.	Analyzing
Knowledge application	After reading Parts 1 and 2: Prepare notes for an in-class debate about ecotourism and conservation in Gandoca-Manzanillo.	Applying
Pre-reading	After reflecting on the first half of the novel and listening to an interview with Rossi, make predictions about what will happen in the remainder of the novel.	Experiencing and analyzing
Initial reading	Parts 3–5: Read with a focus on the comprehension question assigned to the group.	Experiencing
Detailed reading	Parts 3–5: Revisit key passages; present the assigned question.	Experiencing
Critical reading	With selections of Parts 3–5: Complete a collaborative text annotation activity; use the hashtags #GandocaVerde and #DesarrolloGandoca to summarize the arguments for and against the development of the wildlife refuge; include an explanation of why the character holds this belief; read peers' annotations and respond with a counterargument.	Analyzing and experiencing
Knowledge application	Read model Spanish-language petitions from Change.org; complete a collaborative text annotation activity to identify meaning-making strategies for writing a convincing petition; assume the role of one of the characters in the novel and write a petition to argue for/against the development of the wildlife refuge.	Conceptualizing and applying

the novel establishes clear connections between fighting for environmental justice, dismantling the patriarchy, and ending white supremacy (Rhoden & Kearns, 2012, p. 168), students are not likely to make these connections unless they engage in analyzing activities that require them to frame textual meaning critically. If students are going to use their language proficiency and content knowledge to participate in real-world discourse communities, their education needs to illuminate the interests, beliefs, and power structures that collide in sustainability debates. Ensuring that students engage in the knowledge process of analyzing is one way to develop this aspect of sustainability literacies and prepare them to take on the systems of power that prevent real progress towards more sustainable futures.

We also know that sustainability literacies entail the ability to strategically negotiate meaning and build relationships with groups of people who may not share our beliefs and values. For this reason, the unit incorporated frequent applying activities that required students to design meaning in different settings. They participated in oral debates after reading the first half of the novel and, at the end of the unit, students applied their new knowledge about ecotourism and wrote a petition intended to persuade Costa Rican communities that have a stake in the debate about the wildlife refuge. Two different text annotation activities prepared them for this real-world task. First, students did a critical re-reading of selections from the novel in order to strategize the content of their petitions; they used Perusall, a social reading platform integrated into the university's learning management system, to review the arguments for and against the development of the wildlife refuge. They tagged each argument with a hashtag (#GandocaVerde and #DesarrolloGandoca), analyzed the motives and the purposes of the character making the argument, and then brainstormed possible counterarguments. After the collaborative annotation activity had been completed, students could use the hashtags to locate relevant arguments in the text quickly and outline their petitions.

A second textual annotation task prepared students to creatively transform the arguments presented in *La loca de Gandoca* (which emerged in conversational exchanges among characters) and reframe them in a petition. After reading two model petitions from Change.org, the class collaborated on Perusall to conceptualize the meaning-making tools that make petitions persuasive. Teacher-initiated annotations drew students' attention to key language forms and textual conventions, and students responded with their hypotheses about the effect that each choice had on meaning-making. Finally, in-class activities solidified the reading–speaking–writing link necessary for advanced writing development (Byrnes et al., 2010, p. 55).

When writing their own petitions, students transformed the content of *La loca de Gandoca* and the genre conventions of the model petitions they had studied. In a pedagogy of multiliteracies, students may find

> meaning-making resources (…) in representational objects, patterned in familiar and thus recognizable ways. However, these objects are

reworked. Meaning makers do not simply use what they have been given: they are fully makers and remakers of signs and transformers of meaning.

(Cope & Kalantzis, 2009, p. 175)

Students drafted their petitions and then completed a peer review activity that framed writing in this way—as a dynamic, purposeful, and context-bound act of meaning design (Turpin, 2019). As the instructor, I also commented on each rough draft, intentionally focusing my feedback on the meaning-making choices that helped or hindered students' ability "to exercise [their] agency with an identity that is recognized by others in the community" (Swaffar & Arens, 2005, p. 2) and persuade their intended audience. Students addressed issues with content and organization in their peer review groups, and then, as a whole class, we addressed the linguistic issues that would detract from the petition's effectiveness (Byrnes et al., 2010, p. 137). For example, students struggled to consistently and appropriately refer to the reader (mixing *tú*, *usted*, and *ustedes*), select sentence structures that indicate whom to blame for destructive actions (contrasting of active and passive voice), and delineate which aspects of ecotourism and sustainable development exist, and which do not (contrasting of indicative and subjunctive moods). Students used the feedback they had received from peers and the instructor to revise their meaning-making choices and prepare a final version. This genre-based assessment task aligned with the unit objectives because it held students accountable for the content knowledge they had gained about ecotourism and required them to put knowledge into practice in a real-world application of multiliteracies.

Conclusion

The multiliteracies framework can be used in any content-based foreign language class to scaffold students' development of sustainability literacies. In this context specifically, it affords us the theoretical coordinates and practical procedures that we need to prepare students to take active roles in discourse communities committed to sustainability. By theorizing reading and writing as an act of meaning design, it emphasizes the sociocultural context in which texts are produced and interpreted. Instructors can leverage the context-driven and interactive nature of reading to prepare procedural and instructional scaffolding that supports readers through interpreting difficult texts. Furthermore, instructors can use the KP framework to ensure that students move beyond their own experiences with sustainability and have the opportunity to interpret and create texts in which content knowledge, linguistic competence, and sociocultural context intersect.

Applying multiliteracies pedagogy and sustainability education to our language classes may not be easy if we are bound by the confines of a textbook that emphasizes grammatical content and lacks meaningfully integrated texts (Allen & Paesani, 2010, p. 125; Warner & Dupuy, 2018, p. 122). Whether imposed by a textbook, program goals, or university requirements, external pressure can limit the time that instructors and curriculum designers can devote to complex text-based thinking. Developing students' multiple literacies requires extensive time and committed engagement—both of which diminished with the sudden onset of emergency remote instruction at the start of the COVID-19 pandemic. As we transition back to in-person instruction, practitioners are encouraged to leave ample time for multiliteracies instruction,[4] which offers practical and impactful strategies to foreign language educators who aim to empower the next generation of sustainability activists.

Notes

1 The unit was designed to be five weeks long, with 50-minute classes meeting three times per week.
2 Anacristina Rossi is a Costa Rican writer and environmental activist. She was born in San José in 1952 and continues to publish today.
3 Some activities involve multiple knowledge processes. In a prediction activity, students draw upon their personal experience and background knowledge (*experiencing*) to come up with a list of possibilities, then use their knowledge of how texts create meaning in a given context (*analyzing*) to select the most likely possibility and formulate their prediction.
4 The use of social reading platforms (such as Perusall) promotes active and collaborative meaning-making practices. The Center for Advanced Research on Language Acquisition (CARLA) also offers online resources and summer institutes for practitioners who want to implement multiliteracies pedagogy at all levels of post-secondary instruction.

References

Allen, H. W., & Paesani, K. (2010). Exploring the feasibility of a pedagogy of multiliteracies in introductory foreign language courses. *L2 Journal, 2*(1), 119–142. https://doi.org/10.5070/L2219064

Brinton, D., & Snow, M. A. (2017). The evolving architecture of content-based instruction. In M. A. Snow & D. Brinton (Eds.), *The content-based classroom: New perspectives on integrating language and content* (2nd ed., pp. 2–20). Ann Arbor, MI: University of Michigan Press ELT.

Byrnes, H., Maxim, H. H., & Norris, J. M. (2010). Realizing advanced foreign language writing development in collegiate education: Curricular design, pedagogy, assessment. *The Modern Language Journal, 94*, i–235. Retrieved May 1, 2021, from http://www.jstor.org/stable/40985261

Cammarata, L. (Ed.). (2016a). *Content-based foreign language teaching: Curriculum and pedagogy for developing advanced thinking and literacy skills* (Reprint edition). New York, NY: Routledge.

Cammarata, L. (2016b). Foreign language education and the development of inquiry-driven language programs: Key challenges and curricular planning strategies. In L. Cammarata (Ed.), *Content-based foreign language teaching: Curriculum and pedagogy for developing advanced thinking and literacy skills* (Reprint edition, pp. 123–146). New York, NY: Routledge.

Center for Advanced Research on Language Acquisition. (2021). Foreign Language Literacies. https://carla.umn.edu/literacies/index.html

Cope, B., & Kalantzis, M. (2009). "Multiliteracies": New literacies, new learning. *Pedagogies: An International Journal, 4*(3), 164–195. https://doi.org/10.1080/15544800903076044

de la Fuente, M. J. (2019). Stepping out of the language box: College Spanish and sustainability. In C. Melin (Ed.), *Foreign language teaching and the environment: Theory, curricula, institutional structures* (pp. 130–145). New York, NY: The Modern Language Association of America.

Fterniati, A. (2010). Literacy pedagogy and multiliteracies in Greek elementary school language arts. *The International Journal of Learning, 3*, 319–350. Retrieved December 9, 2020, from https://www.researchgate.net/publication/233387450_Literacy_Pedagogy_and_Multiliteracies_in_Greek_Elementary_School_Language_Arts

Genesee, F., & Lindholm-Leary, K. (2013). Two case studies of content-based language education. *Journal of Immersion and Content-Based Language Education, 1*(1), 3–33. https://doi.org/10.1075/jicb.1.1.02gen

Kalantzis, M., Cope, B., Chan, E., & Dalley-Trim, L. (2016). *Literacies* (2nd ed.). Port Melbourne, Australia: Cambridge University Press.

Kautz, E. A. (2016). Exploring environmental and sustainability issues in the intermediate-level foreign language curriculum. In L. Cammarata (Ed.), *Content-based foreign language teaching: Curriculum and pedagogy for developing advanced thinking and literacy skills* (Reprint edition, pp. 234–249). New York, NY: Routledge.

Kearns, S. (1998). Otra cara de Costa Rica a través de un testimonio ecofeminista. *Hispanic Journal, 19*(2), 313–339. Retrieved December 14, 2020, from http://www.jstor.org/stable/44284571

Kern, R. (2000). *Literacy and language teaching* (1st ed.). Oxford, United Kingdom: Oxford University Press.

Kern, R. (2003). Literacy as the new organizing principle for foreign language education. In P. C. Patrikis (Ed.), *Reading between the lines: Perspectives on foreign language literacy* (pp. 49–50). New Haven, CT: Yale University Press.

Kern, R. (2008). Making connections through texts in language teaching. *Language Teaching, 41*(3), 367–387. https://doi.org/10.1017/S0261444808005053

Melin, C. A. (2013). Climate change: A "green" approach to teaching contemporary Germany. *Die Unterrichtspraxis/Teaching German, 46*(2), 185–199. https://doi.org/10.1111/tger.10140

Melin, C. A. (Ed.). (2019). *Foreign language teaching and the environment: Theory, curricula, institutional structures.* New York, NY: The Modern Language Association of America.

Menke, M. R., & Paesani, K. (2019). Analyzing foreign language instructional materials through the lens of the multiliteracies framework. *Language, Culture and Curriculum, 32*(1), 1–16. https://doi.org/10.1080/07908318.2018.1461898

Met, M. (1999, January). Content-based instruction: Defining terms, making decisions. *NLFC Reports.* https://carla.umn.edu/cobaltt/modules/principles/decisions.pdf

Modern Language Association. (2007). Foreign languages and higher education: New structures for a changed world. Ad Hoc Committee on Foreign Languages. *Profession*, 234–245.

Paesani, K., Allen, H. W., & Dupuy, B. (2016). *A multiliteracies framework for collegiate foreign language teaching* (1st ed.). Boston, MA: Pearson.

Rhoden, L. B., & Kearns, S. (2012). Questioning modernity, affirming resilience: Eco-pedagogies for Anacristina Rossi's La loca de Gandoca and Homero Aridjis's ¿En quién piensas cuando haces el amor? *Review: Literature and Arts of the Americas, 45*(2), 167–174. https://doi.org/10.1080/08905762.2012.719767

Rossi, A. (2009). *La Loca de Gandoca* (Segunda edición). San José, Costa Rica: Legado.

Rowland, L., Canning, N., Faulhaber, D., Lingle, W., & Redgrave, A. (2014). A multiliteracies approach to materials analysis. *Language, Culture and Curriculum, 27*(2), 136–150. https://doi.org/10.1080/07908318.2014.927883

Snow, M. A., Met, M., & Genesee, F. (1989). A conceptual framework for the integration of language and content in second/foreign language instruction. *TESOL Quarterly, 23*(2), 201–217. https://doi.org/10.2307/3587333

Stibbe, A., & Luna, H. (2009). Introduction. In A. Stibbe (Ed.), *The handbook of sustainability literacy: Skills for a changing world* (pp. 9–16). Cambridge, United Kingdom: UIT Cambridge Ltd.

Stoller, F. L. (2004). Content-based instruction: Perspectives on curriculum planning. *Annual Review of Applied Linguistics, 24*, 261–283. https://doi.org/10.1017/S0267190504000108

Swaffar, J., & Arens, K. (2005). *Remapping the foreign language curriculum: An approach through multiple literacies.* New York, NY: Modern Language Association of America.

ter Horst, E. E., & Pearce, J. M. (2010). Foreign languages and sustainability: Addressing the connections, communities, and comparisons Standards in higher education. *Foreign Language Annals, 43*(3), 365–383. https://doi.org/10.1111/j.1944-9720.2010.01088.x

The New London Group. (1996). A pedagogy of multiliteracies: Designing social futures. *Harvard Educational Review, 66*(1), 60–92.

Turpin, K. M. (2019). Training foreign language learners to be peer responders: A multiliteracies approach. *L2 Journal, 11*(1), 35–60. https://doi.org/10.5070/L211140673

Warner, C., & Dupuy, B. (2018). Moving toward multiliteracies in foreign language teaching: Past and present perspectives… and beyond. *Foreign Language Annals, 51*(1), 116–128. https://doi.org/10.1111/flan.12316

Wiggins, G., & McTighe, J. (2005). *Understanding by design* (2nd Expanded). Alexandria, VA: Association for Supervision & Curriculum Development.

3 Project-Based Language Learning

Addressing Cultural and Linguistic Diversity Issues in Climate Action

Megan M. Ferry

Introduction

Now entering the third decade of the 21st century, we are experiencing the deleterious effects of climate change and widening disparities across class, race, and gender, not to mention a fractured sense of the human commons amid a devastating global health pandemic. Competing social values and concepts of reality have become justifications for increased hostility toward other humans and the natural world. The way we speak about these issues shapes our conceptual understanding of the problems at hand. Halliday (2006) states that the "categories and concepts of our material existence are not 'given' to us prior to their expression in language. Rather, they are construed by language" (p. 179). Chawla (2006) demonstrates how specific languages conceive of human and environmental ecosystems from radically different perspectives.[1] Eppelsheimer et al. (2014) identify monolingualism and monocultural productions of knowledge as contributors to some of our current conditions because monolingualism's singular perspective fails to conceptualize the entirety of our moment and thus, cannot "do justice to the complexity of the relationships between mankind and the nonhuman environment" (p. 64). This chapter considers how human interconnectedness, diversity, and interdependence with the natural environment serve as resilient forces for charting our way forward out of our current crises. Furthermore, I suggest that college language educators can contribute to addressing these human and ecological sustainability crises through experiential learning approaches such as project-based language learning (PBLL). For this, we need to commit to adapting our language teaching to engage students intellectually at all levels, from beginning to advanced (Halbach, 2014).

I argue in this chapter that PBLL is a form of content-based language instruction (CBI) that integrates subject learning with language acquisition by having students examine open-ended, challenging issues and work toward a goal: Creating a public product for an authentic audience. This pedagogical approach offers a creative way to analyze environmental and social justice topics in the language classroom and engage deeply with them. A project-based framework illuminates constructive and proactive approaches to addressing diversity in the curriculum. "[E]ducation is fundamentally political

DOI: 10.4324/9781003080183-5

since it is constantly involved in the (re)production of social and cultural inequalities (both within and between nations), and of particular forms of culture and knowledge" (Pennycook, 1989, pp. 590–591).[2] The multilingual perspectives that language programs afford are pertinent to the current discussion of racial, economic, and limited English proficiency (LEP) disparities in US education, which, in themselves, speak to a monocultural viewpoint of racial and class privilege (Grawe, 2018; Jaschik, 2018). Thus, it is incumbent upon us to recognize and assert language learning's integrated role in tackling these critical problems we are facing.

A PBLL-informed curriculum can play a significant and central role in students' intellectual and cultural development in that its content-infused instruction "makes the same academic, cognitive and social demands on students as it does for other disciplines and displays the same potential for an array of positive outcomes" (Moritoshi, 2014, p. 12; Stoller, 2006). Furthermore, a PBLL-informed classroom allows for the alignment of the institutional goals of equity, access, and inclusion with the macro-policies of national- and international-level organizations. It has the potential to upset the hierarchy of privilege that has been foundational to US education and to engage language educators as political defenders of democracy by using the language classroom to critically examine our worldviews and to confront "our own preconceptions" (Brydon-Miller, 2006, p. 50). As language educators, our ability to recognize and convey the importance and salience of linguistic and cultural diversity precisely at a moment when the future of our planet and human existence is threatened, has the potential to centralize language learning as an educational imperative. Research has proven that a CBI approach to language learning enriches students' intellectual development, critical thinking, and problem-solving skills, as well as increases student motivation to continue studying a second or third language by integrating it with a subject matter in the language classroom (Cammarata et al., 2016).

This chapter provides a rationale for how PBLL augments the benefits of CBI by proactively engaging students with real-world problems and creating a product that is shared with people beyond the classroom. Students enhance their communication and collaborative teamwork skills, and engage in self-reflection and critical thinking while working on open-ended, challenging issues with authentic tools and activities relevant to their lives. They produce a public product through collaborative teamwork. PBLL is a high-impact learning opportunity that deepens intercultural and linguistic competencies (Cox & Montgomery, 2019), incorporates active learning and personal accountability (Slater & Beckett, 2019), and has the potential, I argue, to transfer student language learning from a dependency on classes and teachers as drivers of education to the recognition that language is a vehicle of students' own making, and a higher degree of autonomy in the language learning process fosters lifelong learning. Moreover, PBLL in the language classroom deepens students' awareness that second (or third) language learning compels us to think differently, and thus, in seeing the problem from a different perspective, we are better able to see and solve complex sustainability problems, such as social justice or climate change.

In the following pages, I outline the core components of PBLL, review evidence from college-level second language (L2) classrooms of some of the benefits of this approach, and then discuss additional research findings that show how a PBLL framework aligns with the needs of college students in the 21st Century. Next, I address how a PBLL-informed L2 curriculum can shed new insights on sustainability challenges, such as climate change. Lastly, I include a brief overview of a college-level, third-year Chinese language course on climate action that incorporated PBLL to consider the multilingual and multicultural dimensions of climate change.

What Is Project-Based Language Learning (PBLL)?

PBLL is an extension of project-based learning (PBL), a pedagogical approach focused on engaging students in exploring meaningful, real-world issues and having them seek solutions to intellectually challenging problems that have real-world consequences. PBLL extends from John Dewey's philosophy of experiential learning and William Heard Kilpatrick's 'Project Method' at the beginning of the 20th century (Peterson, 2012). Formed at a historical moment of conflict seemingly similar to the one we face today (corporate interests in standardizing education, a global pandemic, and ideological and class division), this method emphasized critical reasoning, a social (rather than artificial) context for learning, and a focus on the problem instead of information for information's sake or rote memorization. This kind of learning offered a counterbalance to the standardized education of the time and might be considered the beginning of student-centered models designed to democratize education access and achievement. Although it has grown in and out of favor in education throughout the 20th century, riding political and societal changes in expectations for learning and outcomes, major proponents for standardizing the quality and instruction of PBL come from the Buck Institute for Education (now renamed PBLWorks), founded in 1987 and HQPBL.org. Both organizations focus on bringing 21st Century Skills into the classroom.[3] Science and English as a Second Language (ESL) programs have been using this method for over 40 years, and published research literature on PBLL increased in the 1990s.

Stoller (2006) established a theoretical model for defining project work in the language classroom. More recently, Stoller and Myers (2020) have created a five-stage framework guide to help language teachers. The National Foreign Language Resource Center (NFLRC) at the University of Hawai'i at Manoa has played a central role in making PBLL accessible to language educators, offering summer and online training and badge certification in PBLL as well as frequent webinars on aspects of PBLL.[4] They currently use the international standards and principles of High-Quality Project-Based Learning (HQPBL), a collaborative effort by PBLWorks, the NFLRC, and other organizations to simplify, standardize, and deepen PBL as a pedagogical practice (Larmer,

2018). HQPBL (2018) criteria are similar to those outlined by PBLWorks and identified in the PBLL literature by Stoller (2006), yet add explicit direction for teamwork and project management, "When students truly collaborate, they are contributing individual voices, talents, and skills to a shared piece of work, while respecting the contributions of others" (HQPBL.org, 2018). Project management is about "using time, task, and tools efficiently" and may incorporate design thinking (HQPBL.org, 2018). Scaffolding is also a crucial element in a PBLL setting.

The above-named entities converge on the idea that a quality project (1) is about a challenging or engaging problem or question that is meaningful to students, (2) is inquiry-based, having students seek information or investigate the problem or question in greater depth and detail, (3) is authentic in that these are real-world problems or questions that people are facing, (4) gives students a voice and choice in the project process and outcome to generate students' sense of ownership and engagement throughout the project, (5) affords ample opportunities for reflection about content knowledge as well as the project implementation process itself, (6) includes critique and revision through multiple feedback mechanisms, such as peers, course instructor, outside experts, or authentic audience, (7) culminates in a public product that connects the in-classroom learning experience to a real-world audience (PBLWorks, 2019). Projects can vary in length from a week to a whole semester or year and culminate in a public product or presentation. Student outcomes are dependent on high-quality teaching practices, and the NFLRC offers rubrics for each step of the planning process, educational videos, and training. Examples of PBLL curricula can be found on the NFLRC's PBLL Repository.[5] The course I discuss at the end of this chapter also originated from the NFLRC PBLL online training.

As a framework, project-based work aligns with best practices and research on high-impact learning, which emphasizes designing curriculum around issues that speak to "enduring understanding" and "essential questions" (McTighe & Wiggins, 2012). Students have the opportunity to "think deeply about complex topics while developing proficiency in the target language (Cammarata et al., 2016, p. 10) because they get to transfer their knowledge to an unfamiliar setting and because the approach "explicitly prioritizes the learning process" as well as the final product (Park & Hiver, 2017, p. 50). Moreover, Cox and Montgomery (2019) found that the authentic nature of the project, its sustained inquiry, as well as students' express involvement in decision-making about the project content, align with the objective of optimal 'flow,' when students are so engrossed in their activities that they lose track of time.

PBLL and Meeting the Needs of College-Level Language Learners

Recognizing the complex challenges that we are facing in the 21st century, many organizations supporting higher education, such as the Association of

American Colleges and Universities, published a call for significant changes in how US colleges deliver education in order to face the realities of our collective future (Association of American Colleges and Universities, 2007). The American Council on the Teaching of Foreign Languages (ACTFL) revised its standards to reflect World-Readiness Standards (National Standards Collaborative Board, 2015).[6] The Modern Languages Association (MLA) called for college and university language departments to restructure their curricula with the goal of promoting translingual and transcultural competencies and strengthen the role of language study in education (McGinnis, 2018; MLA, 2007). Most college-level language majors lack professional-level proficiency when they graduate (ACTFL, 2015; Malone et al., 2004). Similarly, leaders across multiple disciplinary fields have called for the integration of foreign language learning with other academic disciplines (Domcekova, 2010; Jackson & Malone, 2009) and the importance of global competency in US education (Organization for Economic Co-operation and Development & Asia Society, 2018). In short, the status quo in foreign language education at the collegiate level—the bifurcation of language and culture, an overemphasis on national literature, and insular focus on language learning for its own sake—is no longer acceptable. Not only do approaches to linguistic and cultural acquisition have to change, we as language educators must also change our mindset and adapt to our new realities. Yet, many college L2 classrooms are still stuck in a "conservative 'rut'" because they treat language as an academic subject, let grammar-driven textbooks determine the curriculum, and construct language input and output exercises with the unchanging cadence of the introduction, practice, and production of forms. Language programs that follow this model are missing opportunities to trigger deeper, long-lasting language and content learning.

There are many examples in the second language learning research and pedagogy literature of the effectiveness of PBLL (Beckett & Miller, 2006; Gras-Velázquez, 2020). Slater et al. (2006) define PBLL "as a social practice [that] requires language and content learning through planning, researching (empirical and/or document), analyzing and synthesizing data, and reflecting on the process and product orally and/or in writing by comparing, contrasting, and justifying alternatives" (p. 242). They studied student and teacher perceptions of PBLL learning outcomes to determine ways formative assessment of both linguistic and knowledge acquisition better reflect salient achievements. Moritoshi (2014) researched the cognitive benefits and acquired social skills of combined language and content learning in an English as a second language class for first-year Japanese college students. The students were tasked with a PBL-informed project about measuring college students' healthy lifestyles. The project extended textbook learning by teaching students the social skills for forming teams and negotiating a common understanding of healthy lifestyle. The project extended cognitive skills by having students design a survey, field test it, and improve upon it before then analyzing the data and presenting their results. The teacher served as a coach and facilitator to motivate students through scaffolded learning activities that promoted access to crucial knowledge and understanding. The teacher plays an important role in promoting a collaborative community and learner independence, which allows students to navigate open-ended inquiries and build skills, such as time management, decision-making, and collaborative dialogue—all aspects that enhance learner cognition. Kelsen (2018), in a class with 63 undergraduate learners of business English in Taiwan examined how much L2 learners used the target language during the project's phases. He observed a positive increase of the spoken target language on performance tasks for a simulated project on investment, which led to increased team engagement. However, Kelsen did note that students transitioning from high school to college had greater difficulty with the learner autonomy required in a PBLL setting, therefore suggesting the use of mixed college-level teams. Taken together, research on PBLL indicates that it offers significant and measurable benefits to L2 learning outcomes.

The rapid rate of change in our societies and environments indicates that our teaching must accommodate the new demands students face. PBLL helps socialize non-mainstream students into the classroom (Beckett, 2006), but it can help socialize students to a larger, general education community cognizant of the intercultural and multilingual realities of our social and academic lives. The language classroom provides opportunities to explore the constructed nature of our realities as mediated through language, in addition to creating intercultural opportunities to contextualize language beyond its linguistic code, and thereby demonstrate how language, culture, and learning, work altogether to construct and interpret meaning. The amalgam of thinking through a real-world problem by addressing the cultural and linguistic assumptions of the problem and its audience, and then reflecting critically about it, are central components of PBLL.

PBLL framework also aligns with ACTFL's World-Readiness Standards (Miller, 2006), even deepening what Magnan (2017) argues are the underutilized Connections and Communities domains. The selection of challenging, real-world problems that students investigate over an extended period allows them to work through the complexity of a problem as part of a team. The project's interdisciplinary nature accommodates students across a wide range of disciplines, thereby affording the opportunity to integrate learning across courses. The public audience of the final product extends in-class learning beyond the school, thus extending the target-language community beyond the classroom perimeter. For example, the collection of PBLL projects discussed in Gras-Velázquez (2020) indicates how practitioners have used PBLL college-level courses to connect students with communities outside of campus, including the local public schools, civic organizations, adult care facilities, and social justice centers.

Environmental Literacy through Second Language (L2) Learning

Integrating L2 study with environmental literacy can resituate foreign language departments in the academy as uniquely able to negotiate the multivocal perspectives and discourses that have been absent in US higher education and climate action (Pearson et al., 2018), as well as meet the culturally inclusive needs of tomorrow's college students. A PBLL-designed pedagogy can engage students in sustained, comparative inquiry about the environment and ecology with the integration of L2 learning so as to encourage them to think and act differently from the perspectives that have brought us to this current climate crisis. There is evidence from linguists and cognitive scientists that linguistic nuance and complexity shape our reality and condition our thinking about problems and processes (Lakoff & Johnson, 1980). Cognitive linguists teach us that language shapes our mindset and behavior, both of which can be altered through language study (Athanasopoulos et al., 2015).

Learning about climate change in another linguistic and cultural context opens the possibility of thinking and acting differently. Moreover, it can offer learners knowledge about cultural behaviors and beliefs and how this particular knowledge can benefit climate action consciousness and initiatives. Furthermore, when we speak a different language, we come to understand a whole community and its history. Equipped with such information, we can change beliefs and behaviors, thus helping pave the way for more diverse and effective climate action strategies. Two examples about language and culture, one from cognitive linguistics and the other from systems thinking (Meadows & Wright, 2008), illustrate where L2 learning can play a leading role in sustainability efforts because it can not only bridge different conceptual systems but also potentially bring about collaborative change.[7]

This first example relates to how we talk about climate change, especially through conceptual metaphors. Language conveys how we understand the world as well as informs how we talk about it. Environmental sustainability discourse is replete with metaphors specific to the English-speaking context that make the abstract concepts more concrete (Larson, 2011). We now talk about climate change instead of 'global warming' and use other words such as 'greenhouse gas' or 'heat-trapping blanket' (Armstrong et al., 2018). Such metaphors can misconstrue the actual problem of climate change, leading one to think that it is only happening elsewhere where it is warm, or that such a problem can be solved as quickly as, say, opening the door of the greenhouse to let out the excessive heat or removing the blanket (Armstrong et al., 2018). While one may understand the meaning of the linguistic terms, they fail to convey the greater complexity of the problem. Such language also sees the issue from a Western-centric perspective, which primarily privileges economic metaphors over justice and ethics in high-level policy documents. Such documents also reflect dichotomous thinking of the issue as "two opposing value systems, a pristine nature, versus economic growth" (Armstrong et al., 2018, p. 39). Climate change itself is an external foe against whom humans battle, thus obscuring human agency as the source of the problem. If the global discussion on sustainability is primarily driven by these policy documents, what other perspectives and pedagogies are being left out?

The second example suggests why our conceptual understanding of the problem is limited. The United Nations Emissions Gap Report states that the "failure to reduce global emissions by 2030 significantly will make it impossible to keep global warming below 1.5°C" (United Nations Environment Programme, 2020). Analysis of why mitigation action has failed thus far has to do with human cognition. We are mainly educated to analyze issues in terms of cause and effect, identifying an actor and result based on observable elements and processes, or based on recognized patterns (Sterman & Sweeney, 2007). However, climate change requires higher-ordered cognition to map the problem holistically (Meadows et al., 2016), which is not frequently taught in school. We, human actors, cannot see, much less undertake, the necessary actions to alter our situation as a result. Meadows and Wright (2008) identify this as a problem of perspective, failing to identify the systemic structure of the problem. In other words, we are one blind man feeling one part of an elephant, ready to declare the truth of what we perceive but missing the larger structural reality (Meadows & Wright, 2008).

The example from cognitive linguistics offered above illustrates how 'selective' perspectives on climate change frame the current discourse and limit climate action strategies. It also constitutes the basis for a newly developed Chinese language course that integrates PBLL pedagogy so students can address the lack of racial and linguistic diversity in US climate action.

'Climate Change and Me: How Culture Informs My Carbon Footprint'

'Climate change and me' is a third-year Chinese language course taught at an upstate New York private liberal arts college that encourages particular citizen participation in climate action. The course consisted of nine students from all college class levels with intermediate-mid to advanced-low Chinese proficiency who were either language majors or minors. Two of them were heritage speakers of Cantonese, and the majority was pursuing a second major in a social science or science field. This trimester-long course satisfies coursework towards Environmental Studies and the Chinese major. In this course, students analyze the role that culture plays in informing one's carbon footprint and compare sustainability efforts in China and the US before taking a deeper dive into finding solutions to the unequal effect climate change has on minorities, working classes, and underrepresented people in the US. Students first measure their carbon footprints and survey their understanding of, and participation in, sustainability efforts, such as recycling, reusing, and reducing consumption. Students in the course need to attend and report on existing *green* clubs, activities, and actions around campus and note who tends to participate. They then study and compare *green* activities by college students in China and compare the US and Chinese nation-level public service announcements and policies. Students are allowed to choose their project product, as long as it aligns environmental sustainability action with a participating Chinese-speaking community on- or off-campus.

With careful scaffolding and sequencing of terms and concepts, students are guided to use their non-native language to negotiate difficult concepts, chart causes and effects of human and industrial consumer behavior, and explain what climate change is. More specifically, students are asked to:

1   define the negative impact of climate change in the global context using specific L2 grammatical resources to express cause and effect, and vocabulary that refers to beneficial and harmful, as well as dangerous, practices of energy consumption
2   illustrate and explain historical and present energy consumption patterns and habits in the US and China (and elsewhere if desired) using specific grammar and vocabulary related to time and energy
3   compare and contrast the cultural differences in energy consumption in the US and China, as well as the reasons behind them, using topic vocabulary and grammar
4   inquire about and report on the awareness of climate change issues in China by interviewing Chinese native speakers, using appropriate language for asking questions, and being receptive to the fact that expressed perspectives may not coincide with the dominant US cultural perspective
5   advocate for the power of the individual to positively combat the effects of climate change by citing examples from individuals in the US and China using the appropriate grammatical and vocabulary related to the topic
6   research and restate the Chinese government's position on climate change and at least one specific policy or action being taken by an organization or the government
7   use systems thinking[8] tools to create feedback loops of their own consumption patterns, work through and debate cause-effect diagrams related to the economy and sustainability, construct empathy maps to better understand the audience for their projects, thus using their linguistic and cultural skills as a vehicle for communication, negotiation, strategy, and information sharing

After acquiring linguistic, cultural, and conceptual understanding of the various perspectives on climate change, students work in teams of 4–5 people, with each team member playing a specific leadership role (equity officer, notetaker, group leader, scheduler), to design a product that will address the issue of lack of racial and linguistic diversity in US climate action. Students reported gaining a greater linguistic and cultural understanding of Chinese, discovered the benefits of team collaboration, better understood their own limitations (how much time investment the project required, or what it was like to engage in decision-making with people who had alternative viewpoints from their own), and were able to connect the project's activities and product to their anticipated careers. These are two examples of final products and students' reflections:

Two Final Products

One team of students planned an afternoon workshop with native Chinese speakers from an ESL after-school program in a local middle school, where Chinese was used to speak with students about recycling and composting. The workshop was so successful that even the non-Chinese students in the ESL program enthusiastically engaged in it.

Another team planned to host a cross-cultural dialogue with Chinese international students on campus. Unfortunately, they planned the event for the eve of the Lunar New Year holiday, when there was likely to be little interest in participation. Despite the setback of a not-so-successful product (it did not meet the assignment criteria to engage with a native-speaking community), the event was attended by several non-Chinese-speaking faculty and students interested in engaging with the team on the topic. The team discovered rich cultural differences, noting that getting busy college students, Chinese or US, to participate in sustainability efforts is difficult because of their busy schedules and self-focused lives. They learned that their Chinese national classmates would most likely not attend any of the sustainability clubs unless climate action participation was for a course grade, preferring to spend extra time on their studies. Another observation this student team made was the different attitudes between the US and Chinese students towards study and time management and how educational structures socialize individuals. The team conjectured that while US student clubs align with grassroots interests from the bottom-up, a hierarchical or top-down education model, such as China's, might inhibit individual action outside proscribed norms.[9]

Students believe that during the course they gain a deeper understanding of climate change and how culture can influence one's carbon footprint. They report an increase in confidence in speaking, reading, and writing in Chinese. They often express more nuanced aspects of learning, such as the importance of communication for mutual understanding, paradoxes in aligning one's actions with sustainability efforts, and the importance of developing skills for consensus-building. Several students note the discrepancy between their large carbon footprint and that of classmates and recognize how this reflects their class privilege. Students come away from the course with a deeper appreciation of how class, race, gender, and culture dynamics inform the complexities and intricacies of team collaboration.

One of the most difficult realizations for some students is the paradox of reconciling their deep interest in China with their international travel to China for study or travel. That is, in their interviews with Chinese international student partners about their carbon footprints in China and the US, the students recognize that while international exchanges benefit everyone, the amount of carbon used to travel back and forth is a drawback in terms of environmental sustainability practices. This is an authentic, real-time problem with no simple answer, which has yielded rich class discussions.

Conclusion

This chapter addressed how the L2 classroom can be a site for deep learning about human and ecological sustainability through PBLL. The high-impact experiences of PBLL can leave an enduring influence on student beliefs, values, and behaviors. As educators, we can facilitate our students' intellectual and emotional growth, but we also have the ability to foster multilingualism and transculturalism as essential competencies for the well-being of our diverse communities. These are investments in curricular change that are long overdue in higher education. CBI can open students to a broader way of thinking about and valuing the power of language and culture to inform and change one's perspectives and behaviors. Using PBLL to address issues of climate change, for example, teaches students that other cultures see their relationship to nature and the environment differently, and the language they use to speak about these relationships has the potential to open a multitude of perspectives for seeking solutions to our current climate and humanitarian crises.[10]

PBLL's communicative and cognitive activities integrate language acquisition and academic content to enhance salient metacognitive benefits that grant students and educators agency in using their knowledge of human linguistic diversity and cultural interdependence to negotiate our current crises. Using a PBLL framework to teach about sustainability requires that students also learn how linguistic nuance and complexity help form our reality and conditions our thinking about problems and processes. We do not just understand an individual when we speak a different language; we come to understand the whole community and its history wherein that individual lives. PBLL pedagogy engages students in sustained, comparative inquiry about the environment with the integration of language learning and complex systems thinking.

The complexity of environmental and human justice issues we face today requires complex thinking or thinking in systems to chart solutions. A PBLL-informed classroom lets us recognize bio- and human diversity in their larger contexts and our interconnectedness beyond the paradigms of monolingualism and single culture systems. To achieve this, we need to educate our students differently, adapting our teaching strategies to reflect scientifically researched practices that challenge our thinking and understanding of the world. A PBLL course about climate change can promote these competencies, vital to our collective, sustainable future. To be sure, creating the conditions for PBLL requires an upfront investment in time, energy, and resources on the part of the language educator and their institution, but curricular resources are widely available to lessen the burden. Language programs that embrace these student-centered curricular and pedagogical approaches may be more likely to demonstrate the powerful learning impact that this approach has on students by articulating goals beyond language learning (such as collaborative work, problem-solving strategies, and critical thinking). In turn, such goals become salient and irrefutable arguments of why L2 learning is an essential element of higher education.

Notes

1 Chawla argues that the English language, for instance, represents reality in fragments because it breaks time into three conceptual frames (past, present, future) and hierarchizes the relationship between humans and nature, whereas Amerindian languages have a more holistic understanding.
2 Pennycook addresses the political aspect of social relationships of power, especially gender, race, and class.
3 21st Century Skills encompass literacies, skills, abilities, and learning dispositions across various fields: Digital, civic, economic, technology, etc., that are deemed essential for the future workforce. These skills include higher-order thinking skills, deeper learning, and deeper self-awareness, as well as collaborative or teamwork skills (Greenhill & American Association of Colleges for Teacher Education, & Partnership for 21st Century Skills, 2010). Other organizations have similar skills lists, such as the American Association of Colleges and Universities, the World Economic Forum, and the National Research Council, to name a few.
4 The NFLRC is one of 16 government-funded resource centers for language education in the United States.
5 https://nflrc.hawaii.edu/pebbles/
6 https://www.actfl.org/resources/world-readiness-standards-learning-languages
7 Systems thinking is a way of viewing the overall structure, patterns, or cycle of a system instead of only discreet aspects or parts of the larger whole.
8 It helps to articulate problems from different perspectives using various methods commonly used in systems thinking, some of which are named here.
9 The second time the course was taught, a team used this information to create a well-researched and designed pamphlet to encourage greater Chinese international student cooperation in our college's non-credit-bearing environmental sustainability clubs and activities, using it as an example of a holistic, liberal education model.
10 See PBLL project on the Japanese concept of *mottainai* (もったいない、勿体無い) (Simpson, 2016).

References

American Council on the Teaching of Foreign Languages (ACTFL). (2015). *ACTFL oral proficiency in the workplace.* Alexandria, VA: ACTFL. https://www.actfl.org/sites/default/files/guidelines/OralProficiencyWorkplacePoster.pdf (accessed March 11, 2021).

Armstrong, A., Krasny, M., & Schuldt, J. (2018). *Communicating climate change.* Ithaca, NY: Cornell University Press.

Association of American Colleges and Universities (AAC&U). (2007). *College learning for the new global century: A report from the national leadership council for liberal education & America's promise.* Washington, DC: AAC&U.

Athanasopoulos, P., Bylund, E., Montero-Melis, G., Damjanovic, L., Schartner, A., Kibbe, A., Riches, N., & Thierry, G. (2015). Two languages, two minds: Flexible cognitive processing driven by language of operation. *Psychological Science, 26*(4), 518–526. doi: 10.1177/0956797614567509

Beckett, G. (2006). Beyond second language acquisition: Secondary school ESL teacher goals and actions for project-based instruction. In G. Beckett & P. Miller (Eds.), *Project-based second and foreign language education: Past, present, and future* (pp. 55–70). Greenwich, CT: Information Age.

Beckett, G., & Miller, P. (Eds.). (2006). *Project-based second and foreign language education: Past, present, and future.* Greenwich, CT: Information Age.

Brydon-Miller, M. (2006). Photovoice and Freirean critical pedagogy: Providing a liberatory theoretical framework to project-based learning in second language education. In G. Beckett & P. Miller (Eds.), *Project-based second and foreign language education* (pp. 41–53). Greenwich, CT: Information Age.

Cammarata, L., Tedick, D., & Osborn, T. (2016). Content-based instruction and curricular reforms: Issues and goals. In L. Cammarata (Ed.), *Content-based foreign language teaching* (pp. 1–21). New York: Routledge.

Chawla, S. (2006). Linguistic and philosophical roots of our environmental crisis. In A. Fill & P. Mühlhäusler (Eds.), *Ecolinguistics reader: Language, ecology and environment* (pp. 115–123). ProQuest Ebook Central. https://ebookcentral.proquest.com

Cox, C., & Montgomery, C. (2019). A study of 21st-century skills and engagement in a university Spanish foreign language classroom. *Foreign Language Annals, 52*(4), 822–849. doi: 10.1111/flan.12426

Domcekova, B. (2010). Science in foreign language education: A response to MLA reports from a liberal arts college Spanish program perspective. *Hispania, 93*(1), 139–143. https://www.jstor.org/stable/25703408

Eppelsheimer, N. Küchler, U., & Melin, M. (2014). Claiming the language ecotone: Translinguality, resilience, and the environmental humanities. *Resilience (Lincoln, Nebraska), 1*(3), 54–68. doi: 10.5250/resilience.1.3.005

Gras-Velázquez. (Ed.). (2020). *Project-based learning in second language acquisition: Building communities of practice in higher education.* New York: Routledge.

Grawe, N. D. (2018). *Demographics and the demand for higher education.* Baltimore, MD: Johns Hopkins University Press.

Greenhill, V., & American Association of Colleges for Teacher Education, & Partnership for 21st, Century Skills. (2010). *21st-century knowledge and skills in educator preparation.* Partnership for 21st Century Skills (P21). https://files.eric.ed.gov/fulltext/ED519336.pdf

Halbach, A. (2014). Teaching (in) the foreign language in a CLIL context: Towards a new approach. In R. Breeze, C. Llamas Saíz, C. Martínez Pasamar & C. Tabernero Sala (Eds.), *Integration of theory and practice in CLIL* (pp. 1–14). Boston, MA: Brill.

Halliday, M. A. K. (2006). New ways of meaning: The challenge to applied linguistics. In A. Fill & P. Mühlhäusler (Eds.), *Ecolinguistics reader: Language, ecology and environment* (pp. 175–202). London and New York: Continuum.

HQPBL.org. (2018). *A framework for high quality project based learning.* https://hqpbl.org/

Jackson, F., & Malone, M. (2009). *Building the foreign language capacity we need: Toward a comprehensive strategy for a national language framework.* Center for Applied Linguistics: https://www.cal.org/resource-center/publications-products/building-foreign-language-capacity

Jaschik, S. (2018, January 8). Are prospective students about to disappear? *Inside Higher Ed.* https://www.insidehighered.com/admissions/article/2018/01/08/new-book-argues-most-colleges-are-about-face-significant-decline

Kelsen, B. (2018). Target language use and performance in project-based language learning (PBLL). *Journal of Asia TEFL, 15*(1), 199–207. http://doi.org/10.18823/asiatefl.2018.15.1.14.199

Lakoff, G., & Johnson, M. (1980). *Metaphors we live by.* Chicago, IL: University of Chicago Press.

Larmer, J. (2018, March 30). *Connecting the new high quality PBL framework & gold standard PBL*. https://www.pblworks.org/blog/connecting-new-high-quality-pbl-framework-gold-standard-pbl

Larson, B. (2011). *Metaphors for environmental sustainability: Redefining our relationship with nature*. New Haven, CT: Yale University Press.

Magnan, S. (2017). The role of the national standards in second/foreign language education. In N. Van Deusen-Scholl & S. May (Eds.), *Second and foreign language education* (pp. 179–195). Cham: Springer. doi: 10.1007/978-3-319–02246-8_16

Malone, M., Rifkin, B., Christian, D., & Johnson, D. (2004). Attaining high levels of proficiency: Challenges for language education in the United States. *Journal for Distinguished Language Studies, 2*, 67. Retrieved from Center for Applied Linguistics https://cal.org/resource-center/resource-archive/digests

McGinnis, S. (2018). The 2007 MLA report "Foreign languages and higher education: New structures for a changed world," ten years later. *ADFL Bulletin, 44*(2), 102–104. doi: 10.1632/adfl.44.2.102

McTighe, J., & Wiggins, G. (2012). *Understanding by design framework: White paper*. ASCD.org. http://www.ascd.org/ASCD/pdf/siteASCD/publications/UbD_WhitePaper0312.pdf

Meadows, D., Sweeney, L., & Mehers, G. (2016). *The climate change playbook*. White River Junction, VT: Chelsea Green Publishing.

Meadows, D., & Wright, D. (2008). *Thinking in systems*. White River Junction, VT: Chelsea Green Publishing.

Miller, P. (2006). Integrating second language standards into project-based instruction. In G. Beckett & P. Miller (Eds.), *Project-based second and foreign language education: Past, present, and future* (pp. 225–241). Greenwich, CT: Information Age.

MLA Ad Hoc Committee on Foreign Languages. (2007). *Foreign languages and higher education: New structures for a changed world*. New York: Modern Languages Association. http://www.mla.org/flreport

Moritoshi, P. (2014). Exemplifying the language and content learning, cognitive and social benefits of project-based language learning through a mini-research project. *Chugokugakuen Journal, 13*, 9–14. https://core.ac.uk/download/pdf/143642633.pdf

National Standards Collaborative Board. (2015). *World-readiness standards for learning languages*. Alexandria, VA: Author. www.actfl.org

Organization for Economic Co-operation and Development (OECD), & Asia Society. (2018). *Teaching for global competence in a rapidly changing world*. OECD. doi: 10.1787/9789264289024-en

Park, H., & Hiver, P. (2017). Profiling and tracing motivational change in project-based L2 learning. *System, 67*, 50–64. doi: 10.1016/j.system.2017.04.013

PBLWorks. (2019). *Gold standard PBL: Essential project design elements*. Retrieved from PBLWorks https://www.pblworks.org/what-is-pbl/gold-standard-project-design

Pearson, A., Schuldt, J., Romero-Canyas, R., Ballew, M., & Larson-Konar, D. (2018). Diverse segments of the US public underestimate the environmental concerns of minority and low-income Americans. *Proceedings of the National Academy of Sciences—PNAS, 115*(49), 12429–12434. doi: 10.1073/pnas.1804698115

Pennycook, A. (1989). The concept of method, interested knowledge, and the politics of language teaching. *TESOL Quarterly, 23*(4), 589–618. doi: 10.2307/3587534

Peterson, B. (2012). Uncovering the progressive past: The origins of project-based learning. *UnBoxed: A Journal of Adult Learning in Schools,* 8. http://gse.hightech-high.org/unboxed/issue8/uncovering_the_progressive_past/

Simpson, J. (2016). *Mottainai for an eco life.* Retrieved from National Foreign Language Resource Center (NFLRC) PBLL Repository https://nflrc.hawaii.edu/pebbles/prototype/172/

Slater, T., & Beckett, G. (2019). *Integrating language, content, technology, and skills development through project-based language learning: Blending frameworks for successful unit planning.* https://lib.dr.iastate.edu/engl_pubs/246

Slater, T., Beckett, G. H., & Aufderhaar, C. (2006). Assessing projects as second language and content learning. In G. Beckett & P. Miller (Eds.), *Project-based second and foreign language education: Past, present, and future* (pp. 241–262). Greenwich, CT: Information Age.

Sterman, J., & Sweeney, L. (2007). Understanding public complacency about climate change: Adults' mental models of climate change violate conservation of matter. *Climatic Change, 80*(3), 213–238. doi: 10.1007/s10584-006-9107-5

Stoller, F., & Meyers, C. (2020). Project-based learning: A five-state framework to guide language Teachers. In A. Gras-Velázquez (Ed.), *Project-based learning in second language acquisition: Building communities of practice in higher education.* (pp. 25–47). New York: Routledge.

Stoller, F. L. (2006). Establishing a theoretical foundation for project-based learning in second and foreign language contexts. In G. Beckett & P. Miller (Eds.), *Project-based second and foreign language education: Past, present, and future* (pp. 19–40). Greenwich, CT: Information Age.

United Nations Environment Programme. (2020). *Emissions gap report 2020.* Nairobi: United Nations Environment Program. https://www.unep.org/emissions-gap-report-2020

4 Problem-Based Pedagogy for the Advanced College Language Curriculum

Becoming a Multicompetent Language User through Sustainability Education

María J. de la Fuente

Introduction

Sustainability is becoming a central part of the educational curricula in many North American colleges and universities. It has become painfully clear that our world has limited natural resources and that we need solutions to live sustainably and ensure a better future for all generations, and higher education has a significant role to play. According to data from the Association for the Advancement of Sustainability in Higher Education (AASHE's Sustainable Campus Index, 2020), institutions are steadily increasing their development and offering of learning opportunities so that students can solve sustainability challenges and "lead society to a sustainable future" (p. 13). This curricular dimension of sustainability education comprises aspects such as the presence and growth of courses and programs with sustainability-related learning goals and educational activities like immersive experiences, sustainable living and learning communities, sustainability-oriented study abroad programs, presentations by experts, exhibits, etc. There is an increasing consensus that every college student should acquire education for sustainable development (ESD) and that such education needs to come from multiple academic disciplines. This chapter is based on the belief that Foreign Language (FL) education needs to be one of them.

The integration of ESD in FL departments (a) constitutes the most significant contribution they can do to society; (b) can ensure the relevance of departments in the institutional sustainability debate (see Cattell & Kleinhans, this volume), (c) is one of the most effective pathways to a curricular model "where content expands to other disciplines with the goal of cross-cultural literacy in the L2" (Swaffar & Arens, 2005), (d) can produce "advanced- to superior-level users of foreign languages in order to meet the growing societal and global demands for multilingualism" (Malovrh & Benati, 2018).

To be sure, an integration of ESD in FL curricula requires the use of constructivist, inquiry-based learning approaches at all levels of instruction—the

DOI: 10.4324/9781003080183-6

same approaches favored by ESD, paired with an L2 research-based pedagogical framework, such as Content-based Instruction (CBI)—FL education where disciplinary content and target language are integrated, thus facilitating content and language learning (Cammarata, 2016; Stoller, 2008). In FL instruction, a constructivist view of learning means the use of specific learner-centered instructional approaches such as project-based language learning, problem-based learning (PBL), or translanguaging learning, among others, which are based on what is known about how an L2 is learned in the classroom setting and how learners can achieve high proficiency levels in the classroom. For example, according to second language acquisition (SLA) research, students need to produce *complex* L2 in order to reach higher levels of L2 proficiency (Brown & Brown, 2014). Therefore, in the upper level of the curricula, we need pedagogical approaches with multiple production-based tasks that require complex language, such as argumentative speaking. Solving sustainability problems through debate and negotiation requires complex modes of communication. Therefore, PBL is one pedagogy that can facilitate a process that, according to SLA research, promotes advanced L2 development.

In what follows, I present PBL as a learner-centered approach to FL education that promotes both high levels of advanced language abilities and sustainability literacy. In particular, I propose the analysis of sustainability problems or *cases* as an effective strategy for college students to develop higher-level competencies as advanced bilinguals (or multilinguals) with cross-cultural sustainability awareness and knowledge. I start by presenting some insights from research on classroom SLA that inform CBI, a necessary framework to achieve both sustainability and advanced L2 literacy. After making a case for sustainable development as critical content, I introduce problem-based pedagogy and sustainability cases, providing examples from a sustainability case used in a Spanish language course—'Environmental and social sustainability in Latin America'[1]—to illustrate how PBI promotes productive learning processes. First, a synopsis of the case follows so that the reader can contextualize the examples.

Synopsis of the Darien Gap Case Study

[Students analyze and evaluate the case and decide on a sustainable solution that takes into account the point of view of the stakeholders]

The Pan-American Highway is a 16,000-mile-long system of roads that begins in Alaska and ends in Tierra del Fuego, Argentina. There is only one point that parts the highway in two, known as the 'Darien Gap,' a 66-mile-long stretch of jungle between Panama and Colombia. Yaviza is the Panamanian town where the highway stops, and Turbo is the Colombian town where it resumes (see Map 4.1).

The Darien Gap contains several habitats: beaches, mountains, swamps, and tropical forests. It is also a biodiversity hotspot that holds unique species, so large extensions are protected. One example is the Darien National Park, a UNESCO World Heritage Site and Biosphere Reserve. Darien helps sequester greenhouse gases such as polluting carbon, and it is part of the Mesoamerican

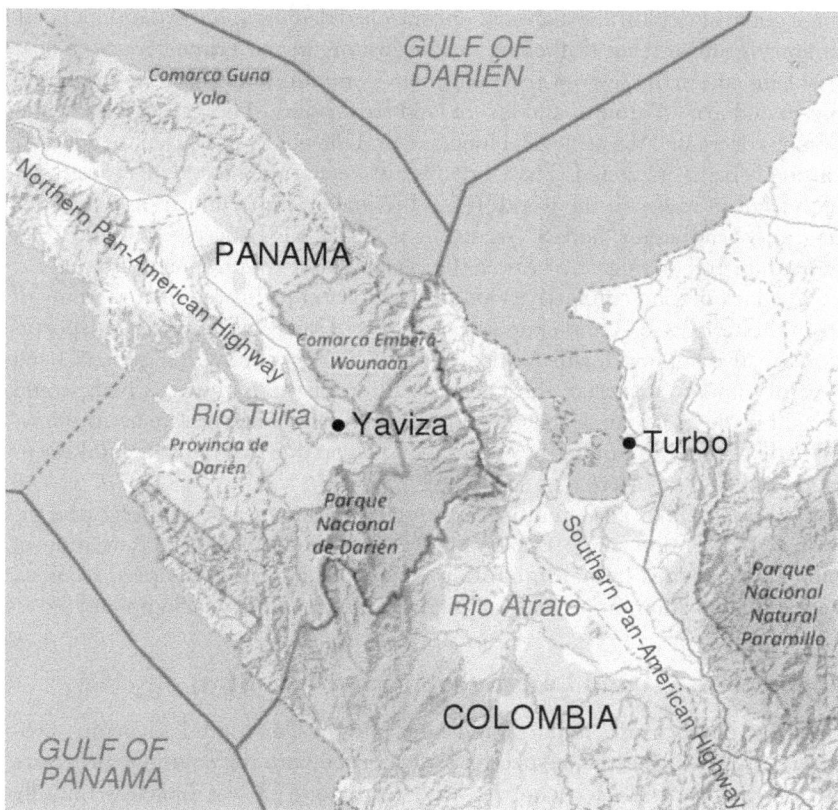

Map 4.1 The Darien Gap, at the border between Colombia and Panama
© CMG Lee / OpenStreetMap contributors / Wikimedia Commons /
CC BY-SA 4.0

Biological Corridor (MBC), which links all the protected areas of Central America. The MBC maintains biological diversity, decreases fragmentation and improves the connectivity of ecosystems, and promotes sustainable production to improve the quality of life of local populations. It also prevents the spread of foot and mouth disease, a viral infection that attacks cattle and pigs, from South America to Central and North America. This tropical rainforest constitutes 20% of Panama, and is home to 35,000 people from ethnic groups. However, control over natural resources is a constant source of conflict. Outside Darien Park, and next to it, is Yaviza, with a population of around 5,000 people. It is a very impoverished area with accessibility difficulties and the highest illiteracy rates in the country. Among the many problems are lack of access to schools and lack of electricity.

The Darien Park is threatened today by many activities in nearby forests and aquifers (the buffer zone): Deforestation due to logging (legal and illegal), timber extraction and trafficking, human-made fires (for livestock ranching), inadequate agricultural expansion, uncontrolled demographic expansion, paramilitary groups, drug trafficking, and other organized crime groups. These problems affect the lives of the indigenous communities: Conflicts with colonizers and armed groups, and loss of land for farming. The forest's fragmentation has decreased the ecological function, and the ability of Darien to sequester carbon is being impacted. The Darien is also one of the most dangerous irregular migrant routes in the world. It has become an important transit point for migrants heading for the US. According to UNICEF, between 2016 and 2020 more than 46,500 migrants crossed the Darien. Many were women and minors.

The lack of land connectivity posits a problem for this region and many of its inhabitants' social and economic progress. Trade and human transportation are also impaired, which impacts the region's economy. There have been several plans to complete the Darien portion of the highway, which would benefit trade in Panama, the neighboring countries, and the Americas. However, there is opposition to the opening of the gap due to environmental, security, and other concerns.

[In this complex case, multiple sustainability dimensions converge and interact. To mention some: Exploitation and erosion of land, conservation of biological systems, human migration, rights of indigenous peoples, security, access to education, improved well-being, and economic development.]

Classroom Second Language Acquisition (SLA)

SLA research provides important insights into the classroom L2 learning process. A psycholinguistic perspective is concerned with the cognitive processes that make it possible to acquire an L2. Although L2 learning was initially considered an input-driven process (Krashen's 1985 Input Hypothesis) and learners needed to be exposed to substantial amounts of *comprehensible input* in the L2, two additional hypotheses assigned a crucial role to learners' L2 production or *output*. The Interaction Hypothesis (Long, 1996) states that interaction in the L2 is responsible for L2 development. Swain's Output Hypothesis (Swain, 1985) claims that comprehensible or *pushed* output (i.e., learners having to produce the L2 during interaction) is needed to develop high levels of communicative (interactional) ability. Within this interactional approach, research has shown that *negotiation* of meaning (i.e., the modifications that learners make to their production due to lack of comprehension from interlocutors), and especially *corrective feedback*, support advanced L2 development (Li, 2010). This view of SLA regards L2 learning as a conscious process that requires attention: Learners need to notice those gaps between their L2 production and the target L2 form (Schmidt, 1990). This research is the framework of highly interactive approaches to L2 instruction: Problem-based and task-based language learning and teaching.

SLA is also concerned with the role that *explicit knowledge* (learners' conscious knowledge of the rules and features) has on *implicit knowledge* (unconscious knowledge accessed in spontaneous production). Acquisition of a second language is commonly associated with implicit knowledge. However, there is evidence that explicit knowledge can facilitate the learning of an L2 (DeKeyser, 1998; Ellis, 1994). For the classroom environment, in particular, this means that the instructor needs to use instructional strategies to draws learners' attention to "some specific grammatical or lexical form in such a way that it helps them either understand it metalinguistically and/or process it in comprehension and/or production so that they can internalize it" (Ellis, 2006, p. 84). The focus on L2 features while engaging in meaningful interaction is an instructional approach called *focus on form* (Doughty & Williams, 1998), and there is evidence that it is more effective when intentional (i.e., instructors design tasks with built-in focus-on-form elements (de la Fuente, 2006; Grim, 2008). This research informs CBI approaches that integrate 'attention to form' activities in the context of meaning-oriented tasks, discovery (of grammar rules) activities, or 'built-in' (grammar) production-based activities (see Sample Activity 4.1).

Sample Activity 4.1

[This is a collaborative, decision-making task from the *analysis* (first) *stage* of the case[2]]

Decide as a group, from the point of view of each of these stakeholders, what the main (a) social, (b) environmental, and (c) economic consequences will be if the road section between Yaviza and Turbo is completed. Then share your opinions with the class.

a   a resident of Yaviza
b   an indigenous leader of Darien
c   an owner of a trading company
d   an environmentalist

Students can carry out the task as a strictly content-based activity. At this point, they have knowledge of the case context and have examined and shared some additional supporting sources. The task involves applying analysis, collaborative work, decision-making, and values thinking competencies (how the same solution can be viewed as sustainable for some, but not for others), so it is not simple. However, it can be carried out with simple L2 production. To avoid this, the task was built with a focus on form around the use of Spanish hypothetical if clauses (if + subjunctive followed by the conditional tense). A focus on form was also added at the reporting stage by asking students to use opinion expressions followed by *que*, creating complex, subordinate sentences.

[Discussion, environmentalist] "*Si terminaran la carretera, se dañaría el corredor biológico de forma irreversible...*" If they finished the road, the biological corridor would be irreversibly damaged.

[Reporting] *"Nos parece que la principal consecuencia ambiental es que se dañaría el corredor biológico de forma irreversible…"* We think that the main environmental consequence is that the biological corridor would be irreversibly damaged).

Another line of research on the SLA processes has examined interactions from the perspective of Sociocultural Theory (Lantolf, 2000), according to which sociocultural experiences mediate (i.e., influence) the development of cognitive processes such as language learning. For the L2 classroom context, this means that learning the language is not just an individual cognitive process, but one significantly mediated by social interaction. Sociocultural Theory uses the construct of Zone of Proximal Development (ZPD), borrowed from psychologist Vygotsky (1986), who believed that effective learning occurs in both the unaided zone (learning on one's own) and the aided zone or ZPD (learning with the help of others). During a classroom interaction, a teacher or another student assists a learner in performing a task through *scaffolding* (Donato, 1994). Through this collaborative dialogue, learners co-construct knowledge, which allows them to learn new L2 features. This theoretical construct (1) reaffirms the need for instructional approaches that actively and purposely engage learners in collaborative tasks that promote meaning and form negotiation, feedback, and restructuring processes—or ZPD work—that will eventually result in the use of advanced structures and vocabulary; and (2) stresses the crucial role of collaborative work, co-inquiry, and meaning negotiation in classes that require discussing and negotiating complex solutions to complex problems.

These are just three examples of how SLA research can inform sustainability-focused, PBL advanced L2 instruction. Very little of this research to date, however, has been directed to the *advanced* L2 learning process (Malovrh & Benati, 2018). The interest in addressing this level of learning only started in the 2000s (Byrnes, 2007a; Byrnes & Maxim, 2004), and in 2007 this research was "remarkably neglected in second language acquisition research" (Byrnes, 2007b, p. 1). Since then, the need for a clearer understanding of what an advanced L2 learner is and what advanced L2 ability looks like continues to be mentioned in articles and volumes devoted to the subject (see Byrnes, 2012; Malovrh & Benati, 2018). Parallel to it, there is a growing interest in advanced FL instruction. Fortunately, this area is poised to grow "as the world continues to demand 'advanced' multilingual actors" (Malovrh & Benati, 2018, p. 5).

Advanced L2 Instruction and Content-Based Pedagogy

A look at traditional college FL curricula tends to reveal an abundance of programs with upper-level courses focusing on *content* and either ignoring or not appropriately addressing language development (Swaffar, 2004). The expectation is that students will arrive at these courses with a sufficient command of the L2 that will enable them to focus on literary and cultural content. The result is students who experience an imperceptible development in L2 abilities. Indeed, the most successful FL programs in terms of language achievement seem to be those that integrate complex content and advanced language instruction in a balanced manner (Leaver & Campbell, 2014). However, in order to do this, faculty needs to be knowledgeable of "curricular and pedagogically appropriate ways to balance content-language instruction and teach toward the highest possible language abilities" (Pfeiffer, 2008, p. 298). CBI, to which Cammarata et al. (2016) refer to as the "engine of FL educational reform" (p. 12), has been, arguably, the most successful way to transform college FL curricula to date. Some instructional strategies that make CBI effective are the use of cooperative learning principles and the information gathering, processing, reporting, and especially, synthesizing in which students engage. By interacting with content in this way, students develop language and content literacy in tandem. Implementing CBI implies attention to L2 development in the upper end of the curriculum.

Over the past ten years, we have seen some inspiring examples of curricular reform initiatives (Brown & Brown, 2014; Cammarata, 2016; Swafar & Urlaub, 2014). For those transforming upper-level FL curricula and courses, defining *advanced language ability* can be a challenging goal. Byrnes (2012) has pointed out that the guidelines and descriptors offered by various agencies (such as ACTFL or the Interagency Language Roundtable—ILR) guide to this day many curricular and instructional decisions. However, these scales of proficiency are simplistic, "tend to be one-shot and often one-sided determinations of language abilities" (p. 510), and make no reference to the context for which the specific L2 ability is intended. The nature of the course content, the specific purposes for which students are learning the L2, the expected competencies they will need, the textual genres, and the types of interactional contexts that they need to be exposed to in order to be competent interlocutors in the global community, are all essential factors in determining the desired advanced L2 outcomes. For example, being able to formulate counterarguments in a clear and organized manner (which requires specific grammatical structures) is part of the profile of an advanced FL learner (see Sample Activity 4.2).

Sample Activity 4.2

[The course involves the use of three textual genres: *expository* (objective presentation and analysis of sustainability problems), *narrative* (historical contexts), and *argumentative* (exposition and exchange of well-founded opinions, arguments,...). This activity is one of several that occur during the decision-making stage (after the analysis stage) in preparation for the final panel debate. It focuses on building *counterarguments*, i.e., to challenge or invalidate opposing arguments. Counterarguments are complex, both language and content-wise.

First, students must understand the other's position because they need to repeat it before giving a counterargument that contains their own, clear position. Counterarguments also require a logical order and coherence. In terms of L2 resources, students' production involves using complex grammar structures: (1) introducing a concession with conjunctions like *aunque* (although) or *a pesar de que* (even though) followed by other's point of view, or (2) introducing other's point of view followed by a contrast expression such as *sin embargo* (however), *no obstante* (nonetheless) and their counterargument.]

After reading the text about the benefits of Darien's opening for Panama's economy, identify the three main benefits mentioned by the Consejo Empresarial Logístico (CEL). Then write three counterarguments of an environmental, social, and economic nature, respectively. Be ready to share them with the class.

In addition to understanding the need for a balanced content-language instruction, it is crucial to acknowledge students' essential bilingual nature: the first language (L1) and the L2 coexist collaboratively in the learners' minds, and "the L1 is inextricably bound up with their knowledge and use of the second" (Cook, 2002, p. 339). This has been evidenced by research that shows how students use both languages productively (Levine, 2011; Scott & de la Fuente, 2008). Extensive observational research has also shown that "exclusive or near-exclusive use of the L2 is rarely encountered in the FL classroom" (de la Fuente & Goldenberg, 2020). Thus, a construct becomes central to reimagining advanced FL instruction: the notion of the FL student as a *multicompetent language learner* (Cook, 2002) who brings her L1 knowledge to the L2 learning process. Our students do not strive to become native L1 users but to develop identities as multilingual human beings. With that understanding, and with our students' awareness of their multicompetent learner status, we can welcome uses of the L1 as a resource when appropriate (for example, as a means for students to engage with background content that exceeds their L2 ability at a given point in time; or to help each other clarify complex concepts or procedural matters, as a scaffolding technique).

Sustainability as Critical Content for College, Advanced FL Instruction

I believe college years should be an educational experience where students acquire sustainability literacy that prepares them to address sustainability problems with the goal of constructing a more sustainable future. Therefore, university curricula, framed under experiential learning approaches, should ensure the acquisition of specific competencies to understand, think, and act to meet the complex sustainability challenges of society (Wiek et al., 2016). As discussions about redefining FL curricula in higher education continue, a new and promising vision has emerged: To combine sustainability studies and FL learning. Examples of successful initiatives can be seen in a recent volume edited by Charlotte Melin (2019) entitled *Foreign Language Teaching and the Environment*, which addresses such a need for transformation by focusing on sustainability and examining "models for teaching language and the environment" (p. 7) and "ways in which the integration of environmental dimensions can become an intentional vehicle for the teaching of intercultural perspectives at every level" (p. 8). Several of the volume's curricular projects are in the area of *environmental humanities* (see Chapter 1, this volume), a new, interdisciplinary field that brings the insights of the humanities to the sustainability field. Others represent innovations where sustainability content was integrated with different types of content-based L2 courses through the use of authentic target-language materials from readily available sources (educational, journalistic, government), used to create activities and tasks to support a sustainability literacy L2 curriculum.

Among the many resources that can inform sustainable development curricular work, a valuable one is the UNESCO's Education for Sustainable Development framework. The framework promotes actions to integrate sustainable development issues in all areas and levels of education and learning to accelerate progress toward sustainable development through its Global Action Programme (UNESCO, 2014). It also constitutes a guide for educational institutions and educators to rethink curricula and pedagogies around sustainability principles and is used by hundreds of universities worldwide.[3] Central to ESD is the Education 2030 Framework for Action[4] (FFA), a document that specifically addresses one of UNESCO's 17 Sustainable Development Goals (SDGs).[5] (UNESCO, 2015). SDG 4 focuses on ensuring quality, inclusive, and equitable education and learning opportunities for everyone. One of the seven FFA areas of action is defined as follows:

> By 2030, ensure that all learners acquire the knowledge and skills needed to promote sustainable development, including, among others, through education for sustainable development and sustainable lifestyles, human rights, gender equality, promotion of a culture of peace and non-violence, global citizenship and appreciation of cultural diversity and culture's contribution to sustainable development.
>
> (UNESCO, 2015, p. 21)

Within this area of action, FL courses and programs can make a unique contribution in higher education if they are willing to engage in transdisciplinary work involving sustainability that also requires engaging with sustainability education goals and teaching approaches.

When undertaking an FL course or curricular initiative involving sustainability, selecting topics should be a relatively easy task; however, operationalizing sustainability-related learning outcomes can be challenging. Wiek et al. (2011) and Wiek et al. (2016) determined six critical competencies for sustainability: System thinking, anticipatory, normative, strategic, collaborative, and problem-solving, which they operationalized as learning objectives (see Table 4.1). Systems thinking is a fundamental competency to analyze problems related to sustainable development, given the multidimensional nature of sustainability. When considering sustainability challenges, three domains or dimensions are at play: The *environmental*, which refers to our ability to live within the means of our natural resources; the

Table 4.1 Sustainability learning competencies (based on Wiek et al., 2016)

	Ability to…
Systems thinking	analyze problems that cross different domains (society, environment, economy) and different scales (from local to global), considering in turn effects, feedback, and other systemic features related to problem-solving frameworks; show knowledge of social groups, their values, preferences, needs, political decisions, and laws; describe how different activities contribute to solve or mitigate a problem.
Anticipatory	understand, anticipate how problems may evolve; understand, evaluate multiple futures; describe the need for prevention, adaptation, mitigation strategies; assess consequences of actions; deal with risks and changes; handle the concepts of temporality (past, present, and future); duration (short or long); uncertainty that allows envisioning possible future situations.
Normative	apply, reconcile, and negotiate sustainable values, principles, and goals; evaluate socio-ecological systems and develop visions of sustainability of these systems; anticipate possible risks derived from action; apply ethical values, like justice, equity, integrity; commit to social and ecological justice, equity and social responsibility; understand and value the diversity of the environment and its implications for justice; understand the social, ecological and economic interconnections of the systems, and the principles of sustainable development.
Strategic	make decisions, design and implement transformable interventions, transitions, and strategies toward sustainability; show tolerance in the face of frustration and ambiguity, which are frequent in sustainability policies and projects; plan and program innovative projects; not hold back against possible external and internal difficulties.
Collaborative	understand and respect others' needs and actions; deal with conflicts in a group; show transcultural understanding; enable and facilitate collaboration and participation in sustainability; face the challenges of sustainability that require collaboration between different disciplines; understand, compare, and evaluate different positions, perspectives, and preferences.
Problem-solving	apply problem-solving frameworks to sustainability problems.

social, concerned with us, humans, our well-being, and our capacity to build sustainable communities; it also includes human rights, security, and other elements of social justice; and the *economic,* which pertains to our ability to use resources responsibly so we can maintain a thriving economy indefinitely. These dimensions interact in complex ways; the ability to consider the balances and trade-offs between those three domains, anticipate futures, and consider their interconnections when making decisions is essential to sustainable development action.

These ESD learning competencies can be used when developing sustainability-based PBL courses and materials (see Table 4.2).

Problem-solving is considered the most crucial competency. It can be said that it integrates all others (Wiek et al., 2016) because it is the ability to apply different problem-solving frameworks to complex sustainability problems and develop viable, inclusive, and equitable solution options that promote

Table 4.2 Darien case study: Sustainability learning outcomes

	Student can…

Systems thinking	— identify the different stakeholders in the case, their values, needs, and interests
	— explain how the environmental, human, and economic dimensions of sustainability interact in the Darien case at the local level
	— explain the cause-effect relationship between the closing of Darien Gap by land and…
	  • increased greenhouse effect gases
	  • improved quality of life of Yaviza inhabitants
	  • … [four other]

Anticipatory	— describe three visions (long-term scenarios) for Darien Park, according to several variables — decide on three measures—prevention, mitigation, and adaptation—that could be taken to alleviate the current threat to Darien Park's biodiversity
Normative	— understand the point of view of all stakeholders — negotiate and identify the trade-offs of different solutions — evaluate a strategy for Darien from the point of view of economic, human, and environmental sustainability — apply ethical values of justice (social and ecological), equity, and social responsibility when making decisions
Strategic	— decide on an innovative plan for a sustainable solution — identify two factors that can affect the success or failure of the plan
Collaborative	— work effectively and cooperatively in a team — understand transcultural differences about values, morals, and perspectives — incorporate conflicting stakeholders' input into activities
Problem-solving	— show basic sustainability problem-solving competence

sustainable development. Problem-solving pedagogy, a widely-used instructional approach for ESD, is also a learner-centered, content-based L2 pedagogy that can promote the development of advanced language abilities. For these reasons, it is the focus of study in this chapter.

Problem-Based Pedagogy and Sustainability Case Studies

PBL is an inquiry-based pedagogical approach. Students are presented with complex, real-world problems that they need to analyze and try to solve. Through engagement in this problem-solving process, they develop knowledge and higher-order thinking skills (Barkley et al., 2005; Duch et al., 2001). When applied to sustainability cases, it can help students develop the competencies outlined in Table 4.1 above. PBL is quite valuable in the FL context, where problems are not solved in isolation but instead integrated within the target L2 societies and cultures in which they occur. As a result, students gain an understanding of other cultures' sustainability values, perspectives, needs, and actions. This is a unique perspective that courses in FL departments can bring to a college education for sustainable development.

PBL use in FL education is relatively recent (Ansarian & Teoh, 2018). Because of its collaborative nature, this learning approach aligns well with the new social approaches to SLA (Block, 2003; Lantolf, 2000), which place the cooperation and co-construction of knowledge during students' interactions at the forefront of the L2 learning process. As mentioned earlier, during PBL, students and instructors make extensive use of the ZPD, primarily through feedback and peer work.

An important consideration when using PBL in content-based L2 learning courses is choosing the appropriate type of problem design. A continuum can be established in terms of the *complexity* of the problem and the *self-directness* of the task (how much independence or guidance are students given to complete all aspects of the task). In 'pure' problem-solving approaches, learners are presented with rather complex problems that may have different solutions (or no solution). They decide how to approach the collaborative work, including finding sources to obtain additional information and perspectives. At the other end of the continuum, the problems are more structured. The instructor provides much of the information needed to solve them, or the solutions may be more obvious or predictable (Jonassen, 2000). Self-directed, unstructured problem-solving cases may not always be an appropriate pedagogical fit for advanced PBL language courses. Courses in the FLs—as opposed to those in other disciplines—present a unique situation in that content must be processed and discussed, to a great extent, via an L2 which, in turn, is also a learning goal. In other words, L2 use mediates both the *language* and the *content* learning processes. This creates additional challenges for the students. In addition, we discussed earlier in the chapter how the complex L2 learning aspect of the course is not entirely an implicit process: The instructor needs to create the opportunities to focus on formal aspects of the L2 during instruction, and this cannot happen in an unstructured environment. This is one main reasons we often find students who took an advanced class where they 'talked a lot' but whose overall L2 proficiency did not make any progress. Other factors that weigh in favor of a more structured approach to problem design are (a) prior knowledge, since sustainability issues—especially in foreign contexts—are usually unknown, and (2) the complexity of the problem itself. All of the above suggest that there is a benefit in leaning toward the structured end of the continuum. Nevertheless, this decision depends on many factors, and the instructor/case designer should decide accordingly[6] (the instructional sequence).

Case Pedagogy

Case studies offer a structured, guided approach to PBL. A case study presents a real and complex situation and often involves a problem that several stakeholders must negotiate (Dunne & Brooks, 2004). In FL courses and curricula working with sustainability cases, the structured format is very valuable (de la Fuente, 2019) because it allows for the use of relatively detailed cases so that students can explore all the perspectives in some depth. We can also integrate role-plays, debates, and panel discussions, which allow for advanced level structures (argumentative) and vocabulary to be the focus of interaction among learners. By structuring a case, we can integrate smaller *tasks* along the way. A task is an activity that students need to carry out using the L2; it has a primary focus on meaning; it engages cognitive processes such as selecting, classifying, or evaluating, and requires a final product (Ellis, 2003, p. 9). For example, sample activities 1 and 2 in this chapter are tasks. Task-based language teaching (TBLT) shares the same SLA principles with PBLL: Both emphasize learners' collaboration and interaction as effective mediators of L2 learning (Bygate et al., 2001; Samuda et al., 2018). Working through a case—such as our Darien case—involves (a) analyzing it and (b) identifying possible solutions and making decisions. Complexity increases from the analysis to the decision-making stage.

Analysis

Analyzing a case starts with understanding its context—geographical, historical, and sociopolitical— the people involved in it, and other essential facts. This first step will help students identify the nature of the sustainability problem and its causes. The problem needs to be clearly formulated and then analyzed, keeping in mind the environmental, social, and economic dimensions of sustainability (see Figure 4.1).

During this stage, asynchronous activities tend to be comprehension-oriented (text or video) and focus on acquiring additional background knowledge by understanding and processing documents, videos, tables, or graphs. For example, identifying the main stakeholders in a case and articulating their position can be done individually (see Sample Activity 4.3)

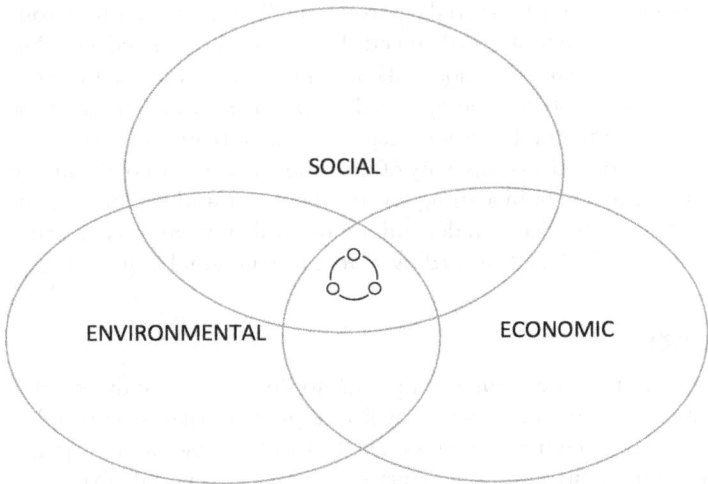

Figure 4.1 The three dimensions of sustainable development

Sample Activity 4.3

1   Read this news report and underline the parts where the points of view of (a) the Yaviza parents, and (b) an environmentalist organization, are expressed. Write each point of view concisely.
2   Watch this video and identify the point of view and one supporting argument of the Panama government. Write the point of view and argument concisely.

In-class tasks target higher-level thinking skills. Many activities engage students in *systems thinking*, i.e., analyzing the multiple aspects of the problem and discovering the connections between them (see Sample Activity 4.4). They also require *normative thinking*: Students' understanding of how other peoples and cultures see the same problem.

Sample Activity 4.4

[The activity has a reporting phase with a built-in focus-on-form feature]

In small groups, analyze the nature of the conflict/problem—the completion of the Darien Gap by land—using the sustainability scheme.

•   Environmentally sustainable = it is compatible with the preservation of biodiversity and ecosystems.
      _____ *y por eso* (and because of that) _____

- Socially sustainable = it has no negative and positive impacts; local communities receive benefits for the development of this activity that improve their living conditions.

  _____ *sin embargo* (however) _____. *Por esa razón* (For this reason)_____

- Economically sustainable = takes into account environmental and social sustainability and is still financially feasible and profitable.

  Aunque (Although) _____, _____. *Por lo tanto* (Therefore) _____

E.g.: Completar el tránsito terrestre puede mejorar la condición de vida de muchos ciudadanos de Yaviza, que no tienen acceso a la electricidad; sin embargo no podemos olvidar el impacto negativo para las poblaciones indígenas del parque. Por esta razón no es socialmente sostenible.

(Completing the land transit can improve the living conditions of many citizens of Yaviza, who do not have access to electricity; nevertheless, we cannot forget the negative impact on the park's indigenous populations. For this reason, it is not socially sustainable.)

Decision-Making

This part of the case involves exploring possible solutions, identifying future sustainable visions, making proposals, and supporting them with arguments. It also requires deciding on a solution (including negotiations and trade-offs) and considering what will happen if no action is taken. As the analysis raises students' awareness of sustainability problems, the decision-making process has to do with taking actions to solve them. For any solution to be truly sustainable, some balance between the three dimensions: (economic, environmental, and social) is needed (see Sample Activity 4.5).

Sample Activity 4.5

[Class is divided into groups, each representing one stakeholder. Stakeholders are representatives of Yaviza parents, Wounan and Emberra leaders, CEOs of trading companies, US government, environmental NGOs, Panama and Colombia governments, and UNICEF. Students need to show all interactions: environmental+social = E+S, economic+environmental = E+E, social+economic = SE. Students will use some of these 'interactions' later on again when they participate in a panel debate.]

Review the stakeholder's position. Using the sustainability scheme, think of possible interactions between the three areas of sustainability that can help support your position.

E.g.: Si terminan la carretera los beneficios económicos y la conectividad serán posible el acceso a mejores escuelas y educación. También un parte de los beneficios económicos pueden ir a la protección ambiental para mitigar los efectos de la apertura del Darién, investir en una sistema mejor de seguridad para eliminar la tala y agricultura ilegale, mejora seguridad y garantizar los pobladores pueden mejorar su vida dentro y cerca el Darién.

(If the road is completed, the economic benefits and connectivity will allow access to better schools and education. Also, part of the economic benefits can go to environmental protection to mitigate the effects of the opening of Darien, invest in a better security system to eliminate illegal logging and agriculture, increase security and guarantee that the population can improve their life in and near Darien.)

In Sample Activity 4.5, students applied *systems thinking* (the interconnectivity of the sustainability dimensions, which is a constant in sustainability problems). They also applied *anticipatory* strategies as they share their vision of the possible. The same strategies are applied in the debates where students discuss and decide the most suitable solutions and the plans that they entail. *Strategic thinking* is also required to plan and anticipate things that could negatively affect their plan (see Sample Activity 4.6).

Sample Activity 4.6

[This is a panel debate in the form of a role-play. All stakeholders are represented. The goal is to find a sustainable solution that considers the needs and views of most, if not all, stakeholders. Before the panel, each group elaborates its plan, develops arguments, and prepares possible counterarguments. In the next class session, the panel discussion takes place, with the instructor acting as moderator.]

Final Remarks

This chapter has illustrated one way to integrate ESD—via problem-based pedagogy—into advanced FL courses. In doing so, we can contribute to what the primary mission of higher education is, or should be: To give students knowledge, competencies, and tools to live sustainably, and to find solutions to the sustainability challenges we face so that future generations can too. Sustainable education as content for teaching FLs is an effort worth pursuing. When taking such a course, students can immediately connect it to one or more different disciplines they are learning about. It is no longer a class they take, so they can travel or talk to family community members who speak a different first language— and these are, for sure, some of the great reasons to become multilingual! Reorienting FL curricula toward environmental and social sustainability issues would make it relevant to many students' academic, personal, and ethical profiles, giving them many more reasons to be in our programs and to *stay* in our programs.

To be sure, this type of curricular and instructional work is not free of challenges. It requires faculty to possess or acquire additional sustainability-related knowledge and, in many cases, a knowledge base in second language learning research, curriculum design, and L2 learning instructional approaches and strategies, such as content-based and problem-based case pedagogy. For the upper-level area of the curriculum, instructors also need a skillful approach to advanced language pedagogy that includes ensuring that students use target structures (instead of strategies like avoidance, for example), using scaffolding techniques, or eliciting target vocabulary while a discussion is taking place (among many other instructional strategies) to maximize target-language learning.

However, all of the above can be approached as an opportunity to explore new disciplines and new forms of teaching and learning, and develop professionally. FL courses and programs need to "step out of the language box" (de la Fuente, 2019) and "transcend disciplinary silos" (Melin, 2019). They need to be on this side of the conversation. As higher education instructors, we have a privileged position that allows us to contribute by giving students the type of FL education that will prepare them for the global challenges ahead.

Notes

1 The course is based on six complex sustainability cases.
2 Actual tasks and activities are more detailed, contain visual material, and offer more procedural guidelines to students. They have been synthesized and translated into English for purposes of space.
3 An example of research on how the ESD framework can be integrated into FL education is Jodoin (2020). The study examined the effectiveness of ESD approaches in English as a foreign language (EFL). Jodoin integrated ESD best practices into a content-based English course in a Japanese university, and compared this course to the same one, offered without the ESD component, concluding that ESD integrated into language teaching correlates positively with environmental behavior change.
4 UNESCO and several other organizations signed this Declaration after the World Education Forum 2015 in Incheon, S. Korea. Participants from 160 countries adopted the Declaration.
5 See https://sdgs.un.org/goals
6 The Darien Gap case follows a structured approach in its design. Students are provided many of the input sources. This approach is preferred for two reasons: (1) the high complexity and context-specific nature of the case, unknown to students; and (2) to ensure appropriate exposure (genres and difficulty levels) to L2 textual and aural genres. Tasks are also structured over several days, from simple to complex (linguistically, thematically, and cognitively).

References

Ansarian, L., & Teoh, M. L. (2018). *Problem-based language learning and teaching: An innovative approach to learn a new language.* Singapore: Springer. https://doi.org/10.1007/978-981-13-0941-0

Association for the Advancement of Sustainability in Higher Education. (2020). *2020 Sustainable campus index.* https://www.aashe.org/sustainable-campus-index/

Barkley, E. F., Cross, K. P., & Major, C. (2005). *Collaborative learning techniques: A handbook for college faculty.* San-Francisco, CA: Jossey-Bass.

Block, D. (2003). *The social turn in second language acquisition.* Washington, DC: Georgetown University Press.

Brown, T., & Brown, J. (Eds.). (2014). *To advanced proficiency and beyond: Theory and methods for developing superior second language ability.* Washington, DC: Georgetown University Press.

Bygate, M., Skehan, P., & Swain, M. (2001). *Researching pedagogical tasks: Second language learning, teaching, and testing.* Harlow, England: Longman. Byrnes, H. (2007a). (Ed.). *Advanced language learning: The contribution of Halliday and Vygotsky.* London: Continuum.

Byrnes, H. (2007b). Locating the advanced learner in theory, research, and educational practice: An introduction. In H. Byrnes, D. Weger-Gunthar & K. Sprang, K. (Eds.), *Educating for advanced foreign language capabilities* (pp. 1–16). Washington, DC: Georgetown University Press.

Byrnes, H. (2012). Advanced language proficiency. In S. Gass & A. Mackey (Eds.), *The Routledge handbook of second language acquisition* (pp. 506–520). New York: Routledge.

Byrnes, H., & Maxim, H. H. (Eds.). (2004). *Advanced foreign language learning: A challenge to college programs.* Boston, MA: Heinle & Heinle.

Cammarata, L. (Ed.). (2016). *Content-based foreign language teaching. Curriculum and pedagogy for developing advanced thinking and literacy skills.* Abingdon: Routledge.

Cammarata, L., Teddick, D., & Osborn, T. (2016). Content-based instruction and curricular reforms: Issues and goals. In L. Cammarata (Ed.), *Content-based foreign language teaching: Curriculum and pedagogy for developing advanced thinking and literacy skills* (pp. 1–21). Abingdon: Routledge.

Cook, V. (Ed.). (2002). *Portraits of the L2 user.* Clevedon: Multilingual Matters.

de la Fuente, M. J. (2006). Classroom vocabulary acquisition: Investigating the role of pedagogical tasks and form-focused instruction. *Language Teaching Research, 10*(3), 263–295. https://doi.org/10.1191/1362168806lr196oa

de la Fuente, M. J. (2019). Stepping out of the language box: Connecting college Spanish to sustainability. In C. Melin (Ed.), *Foreign language teaching and the environment: Theory, curricula, and institutional structures* (pp. 130–145). New York, NY. Modern Language Association.

de la Fuente, M. J., & Goldenberg, C. (2020). Understanding the role of the first language (L1) in instructed second language acquisition (ISLA): Effects of using a principled approach to L1 in the beginner foreign language classroom. *Language Teaching Research.* https://doi.org/10.1177/1362168820921882.

DeKeyser, R. (1998). Beyond focus on form: Cognitive perspectives on learning and practicing second language grammar. In C. Doughty & J. Williams (Eds.), *Focus on form in classroom second language acquisition* (pp. 42–63). Cambridge, England: Cambridge University Press.

Donato, R. (1994). Collective scaffolding in second language learning. In J. Lantolf & G. Appel (Eds.), *Vygotskian approaches to second language research* (pp. 33–56). Norwood, NJ: Ablex.

Doughty, C., & Williams, J. (1998). Pedagogical choices in focus on form. In C. Doughty & J. Williams (Eds.), *Focus on form in classroom second language acquisition* (pp. 197–262). Cambridge, England: Cambridge University Press.

Duch, B. J., Groh, S. E., & Allen, D. E. (2001). *The power of problem-based learning. A practical "how-to" for teaching undergraduate courses in any discipline.* Sterling, VA: Stylus Publishing.

Dunne, D., & Brooks, K. (2004). *Teaching with cases.* Halifax, NS: Society for Teaching and Learning in Higher Education.

Ellis, R. (1994). Explicit/implicit knowledge and language pedagogy. *TESOL Quarterly, 28*(1), 166–172.

Ellis, R. (2003). *Task-based language learning and teaching.* Oxford, UK: Oxford University Press.

Ellis, R. (2006). Current issues in the teaching of grammar: An SLA perspective. *TESOL Quarterly, 40*(1), 83–107. https://doi.org/10.2307/40264512

Grim, F. (2008). Integrating focus on form in L2 content-enriched instruction lessons. *Foreign Language Annals, 41*(2), 321–346. https://doi.org/10.1111/j.1944-9720.2008.tb03295.x

Jodoin, J. J. (2020). Promoting language education for sustainable development: A program effects case study in Japanese higher education. *International Journal of Sustainability in Higher Education, 21*(4), 779–798. https://doi.org/10.1108/IJSHE-09-2019-0258

Jonassen, D. (2000). Toward a design theory of problem solving. *Educational Technology Research and Development, 48*(4), 63–85. https://doi.org/10.1007/BF02300500

Krashen, S. D. (1985). *The input hypothesis.* London: Longman.

Lantolf, J. P. (Ed.). (2000). *Sociocultural theory and second language learning.* Oxford: Oxford University Press.

Leaver, B. L., & Campbell, C. (2014). Experience with higher levels of proficiency. In T. Brown & J. Brown (Eds.), *To advanced proficiency and beyond: Theory and methods for developing superior second language ability* (pp. 3–21). Washington, DC: Georgetown University Press.

Levine, G. (2011). *Code choice in the language classroom.* Clevedon: Multilingual Matters.

Li. (2010). The effectiveness of corrective feedback in SLA: A meta-analysis. *Language Learning, 60*(2), 309–365. https://doi.org/10.1111/j.1467-9922.2010.00561.x

Long, M. H. (1996). The role of the linguistic environment in second language acquisition. In W. C. Ritchie, & T. K. Bhatia (Eds.), *Handbook of second language acquisition* (pp. 59–70). San Diego, CA: Academic Press.

Malovrh, P., & Benati, A. (Eds.). (2018). *The handbook of advanced proficiency in second language acquisition.* Hoboken, NJ: Wiley Blackwell.

Pfeiffer, P. (2008). The discipline of foreign language studies and reforming foreign language education. *The Modern Language Journal, 92*(2), 296–298. https://doi.org/10.1111/j.1540-4781.2007.00719_4.x

Samuda, V., Branden, K., & Bygate, M. (Eds.). (2018). *TBLT as a researched pedagogy.* London: John Benjamins Publishing Co.

Schmidt, R. (1990). The role of consciousness in second language learning. *Applied Linguistics, 11*, 17–46.

Scott, V., & de la Fuente, M. J. (2008). What's the problem? L2 learners' use of the L1 during consciousness-raising, form-focused tasks. *The Modern Language Journal, 92*, 100–113. https://doi.org/10.1111/j.1540-4781.2008.00689.x

Stoller, F. (2008). Content-based instruction. In N. Van Deussen-Scholl & N. H. Homberger (Eds.), *Encyclopedia of language and education (Vol. 4): Second and foreign language education* (pp. 59–70). New York: Springer.

Swaffar, J. (2004). A template for advanced learner tasks: Staging genre reading and cultural literacy through the précis. In H. Byrnes & H. H. Maxim (Eds.),

Advanced foreign language learning: A challenge to college programs (pp. 19–46). Boston, MA: Heinle & Heinle.

Swaffar, J., & Arens, K. (2005). *Remapping the foreign language curriculum: An approach through multiple literacies.* New York: Modern Language Association.

Swafar, J. & Urlaub, P. (2014). *Transforming postsecondary foreign language teaching in the United States.* Dordrecht: Springer.

Swain, M. (1985). Communicative competence: Some roles of comprehensible input and comprehensible output in its development. In S. Gass & C. Madden (Eds.), *Input in second language acquisition* (pp. 235–253). Rowley, MA: Newbury House.

United Nations Educational, Scientific and Cultural Organization. (2014). *Roadmap for implementing the global action programme on education for sustainable development.* Paris: UNESCO.

United Nations Educational, Scientific and Cultural Organization. (2015). *Education 2030. Incheon Declaration and Framework for Action for the implementation of sustainable development goal 4.* Paris: UNESCO.

Vygotsky, L. (1986). *Thought and language.* Cambridge, MA: MIT Press.

Wiek, A. *et al.* (2016). Operationalizing competencies in higher education for sustainable development. In M. Barth, G. Michelsen, M. Rieckmann & I. Thomas (Eds.), *Routledge handbook of higher education for sustainable development.* London and New York: Routledge.

Wiek, A. *et al.* (2011). Key competencies in sustainability: a reference framework for academic program development. *Sustainability Science, 6,* 203–218.

Part 2

Implementing Sustainability-Based Curricular Initiatives in Foreign Language Education

5 Engaging Students with Social, Cultural, and Environmental Sustainability Topics in the Spanish-Speaking World

A Reimagined Beginner Spanish Curriculum

Jorge Méndez Seijas and María Luisa Parra

Introduction: Towards Tailored Content-Based Language Curricula

In the last decade, the field of post-secondary second language (L2) pedagogies has embraced calls to advance learners' language proficiency while engaging them with a host of social, cultural, and environmental issues of local and global relevance (e.g. Barbas-Rhoden, this volume; Dubreil & Thorne, 2019; Klimanova & Hellmich, 2020; Melin, 2019). To address these issues while following contemporary L2 pedagogical frameworks, curriculum developers must overcome significant hurdles, some of which are embodied in commercial language materials. Most commercial materials, particularly introductory Spanish textbooks, are structured around the teaching of grammatical knowledge, oftentimes disregarding evidence-based practices derived from instructed L2 acquisition research that could increase language proficiency (Cubillos, 2014). As it pertains to content, these textbooks tend to present topics that rarely go beyond generic themes such the "the clichéd food, fun, and fiestas [trio]" (Herman, 2007, p. 126); avoid 'controversial' conversations of social relevance (Bori, 2018); and pay little to no attention to meaningful differences in cultural perspectives in the target communities, offering instead a "restricted and restricting tourism discourse and [a] shallow treatment of diversity as multiplicity, not difference" (Kramsch & Vinall, 2015, p. 22). Moreover, the content offered by these materials rarely aligns with students' academic interests, with university-wide initiatives, or with the goals of liberal arts education more broadly.

Developing in-house materials allows programs to tailor their curricula to meet specific programmatic goals while engaging learners with compelling texts and meaningful information about target communities, both

DOI: 10.4324/9781003080183-8

indispensable elements to develop linguistic and cultural competencies (Parra, 2013). Moreover, tailored curricula can be used to more easily integrate and constantly update content that directly responds to the needs and interests of the students. For instance, when restructuring the German language program at Stanford University, a group of professors developed a curriculum in which they tried different ways of adding literary content in their introductory language courses, offering nuanced cultural information that used to be taught exclusively in upper-level courses (Bernhardt & Russell, 1999). In another example, the French language program at Harvard University embarked on a project to not only bring Parisian cultures to their students but also take students on a journey to Paris, to the very places that they are learning about, albeit through virtual reality (Mills et al., 2020). Finally, the School of Foreign Service (SFS) Spanish program at Georgetown University revamped its three-course advanced Spanish sequence to include a series of authentic multimodal texts that deal with pressing issues of social justice (e.g., race and gender discrimination) affecting various minoritized communities in the Spanish-speaking world.[1] These texts help students better understand communities that they might have the opportunity to work with through local internships or sojourns abroad. A commendable conviction guided all these curricular efforts: Besides adding relevant cultural content, these programs believed that language courses should be as intellectually enthralling as any other course offered by their universities.

The development of in-house materials has many other clear benefits, such as enabling the integration of knowledge from instructed L2 acquisition research with socially and culturally responsible content. Additionally, it may facilitate a better alignment of the goals of language programs with those of liberal arts education, thereby contributing to the transformation of students into global citizens committed to enacting positive change. This alignment is only natural given the leverage language programs have in fostering global citizenship by connecting students, linguistically and culturally, with worldviews and social realities different from their own. Enacting a curricular vision such as this requires critical evaluations of any existing materials and content that may be leaving aside complex yet substantial discussions that are crucial for 21st-century education. Selecting the right content and organizing it, however, are challenges to contend with, particularly in introductory language courses.

For the current project, it was determined that content had to (a) be relevant for a diverse pool of students interested in different disciplines within the humanities, social sciences, and economics; (b) include relevant cultural, social, and environmental topics pertaining to a diverse representation (e.g. culturally, racially, socioeconomically) of target language communities; and (c) connect the beginning language program with university-wide and worldwide initiatives. In this context, sustainability-related issues (e.g. pollution, equitable housing, immigration policy) that have historically impacted Spanish-speaking communities represent a vast source of multidisciplinary content that directly connects students' concerns about the world and the future with those of target communities, both locally and around the world. This content—presented through a wide array of multimodal texts (e.g. audiovisual, written, digital, virtual)—was selected and scaffolded in ways that allow for students to engage with the material and understand it for its cultural relevance while critically examining and reflecting upon matters of social justice.

By integrating compelling sustainability-related issues in the new curriculum, this project sought to move beyond dichotomous views of L2 education that place introductory courses and advanced courses at polar opposites of the language-content spectrum (Paesani & Allen, 2012). Instead, this project embraced a content-based instruction (CBI) approach (Cammarata et al., 2016; Lyster, 2007; Snow & Briton, 2017), balancing the teaching of both language and content. The instructional materials and learning experiences developed to analyze the content were designed following the multiliteracies pedagogical framework (Cope & Kalantzis, 2009; Kern, 2000; New London Group, 1996; Paesani et al., 2016), engaging students with tasks that advance their literacy, a meaning-making process that "entails at least a tacit awareness of the relationship between [texts'] contextual conventions and their context of use, and ideally, the ability to reflect critically on those relationships" (Kern, 2000, p. 16).[2]

This curricular project took place in a series of phases that went from the conceptual design to the implementation and evaluation of the new second-semester course.[3] These phases are explained in the following four sections. First, a rationale for the sustainability-based content selected is provided; second, the pedagogical approaches undergirding the curricular decisions are presented and justified; third, the course development process is explained, specifying content, goals, texts, and assessments included in each unit; and fourth, the results from an evaluation consisting of a series of surveys for students and surveys and interviews for instructors, the two primary stakeholder groups, are presented and discussed.

Content and Context

Recognizing the importance of sustainability in protecting the world for future generations, in 2015 the United Nations (UN) adopted its 2030 agenda for sustainable development. This agenda includes 16 goals organized around three main dimensions: social issues (e.g. inclusive and sustainable cities, responsible consumption and production), cultural issues (e.g. safeguarding cultural heritage, intercultural dialogue), and environmental issues (e.g. climate change, desertification). The university where this course was developed has echoed the UN's preoccupation for the future by offering a growing number of seminars, workshops, inter-departmental working groups, and student- and faculty-led initiatives for divestment that have become part of the campus's ongoing conversations regarding sustainability. There is even an office exclusively dedicated to harnessing a more sustainable future, supporting faculty, students, and green labs that contribute to various undertakings through several university-wide endeavors. Because global challenges are always locally rooted, and global change often originates in local ideas, sustainability-based content presents an opportunity to offer students a glocalized learning experience, defined as "curricula consideration and pedagogical framing of local and global community connectedness in relation to social responsibility, justice, and sustainability" (Patel & Lynch, 2013, p. 223).

While the environmental dimension of sustainability contemplated in the UN agenda has been widely incorporated in language curricula (e.g. Melin, 2014, 2019), the social and cultural dimensions of sustainability have not been as widely explored (de la Fuente, 2019). The new second-semester course aimed to integrate all three dimensions in its four units (see Table 5.1) in order to better align the beginning Spanish program "with lifelong learning objectives that extend beyond classroom study and [...] foster the development of globally aware dispositions in our students" (Dubreil & Thorne, 2019, p. 9). Moreover, the organization of the course around a set of well-defined sustainability topics allows students to advance towards intermediate language proficiency and enhance their cultural competencies while establishing clear connections with

Table 5.1 Thematic units and learning goals

Unit	General goals
1 The clothing industry: Sustainable consumption and production	• Demonstrate critical understanding of the clothing industry (production) and its role in environmental and human rights issues/violations. • Demonstrate critical understanding of our responsibility (consumption) pertaining to pollution and human rights violations in the clothing industry.
2 Food as an intangible cultural heritage	• Identify the different features of local cuisines in the Spanish-speaking world. • Demonstrate some understanding of the relevance of these local cuisines as an intangible cultural heritage.
3 Sustainable cities and communities	• Demonstrate critical understanding of some of the sustainability challenges that certain communities in the Spanish-speaking world face in their daily lives. • Compare sustainability indicators and data from international organizations to residents' perspectives from various countries in the Spanish-speaking world.
4 Migrants and the world they shape	• Identify the diversity (e.g. racial, ethnic, political, linguistic) among Latinos/as/@s/xs/es in the US. • Demonstrate critical understanding of some of the inequalities (e.g. access to jobs, quality education) faced by migrants in the US and the Spanish-speaking world.

disciplines such as environmental sciences, sociolinguistics, migration, border and Latinx studies, and history. In what follows, the pedagogical frameworks that allowed for a seamless translation of such complex content into accessible and appropriate class materials and assessments are presented.

Content-Based Instruction and the Multiliteracies Pedagogy

For decades, scholars and professional organizations in the field of L2 education have advocated for language and content to be interwoven at all levels of instruction (MLA report, 2007; National Standards Collaborative Board, 2015; Swaffar & Arens, 2005). This position has been a response to the persistence of a two-tiered system wherein beginner classes focus almost exclusively on language development and upper-level classes focus on content, literary or otherwise. In light of this divide, the Modern Language Association (MLA) Ad Hoc Committee on Foreign Languages recommended in their 2007 report that programs replace this two-tiered structure "with a broader and more coherent curriculum in which language, culture, and literature are taught as a continuous whole" (p. 3). CBI represents a solution to this challenging and elusive language-content articulation dilemma because it is "specifically designed to embed language instruction in the context of content that is meaningful to learners" (Cammarata *et al.*, 2016, p. 20). By promoting a dynamic and balanced interplay between language instruction and content (Lyster, 2007), CBI simultaneously increases students' communicative abilities while enhancing subject-matter expertise, advanced thinking, and critical awareness and disposition (Cammarata, 2016; Sato et al., 2017). All these skills can, in turn, aid in the advancement of translingual and transcultural competence, that is, "the ability to operate between languages [and cultures]" (MLA report, 2007, p. 3). The proven effectiveness of CBI in promoting both content and language learning (Cumming & Lyster, 2016) has resulted in its use in a wide array of educational contexts, from immersion to K-12 to post-secondary education (Cammarata et al., 2016; Lyster, 2017).

To make the new content-based curriculum accessible to students, concrete instructional materials and learning experiences were designed following the *multiliteracies pedagogy* (Cope & Kalantzis, 2009; Kern, 2000; New London Group, 1996), a "socially responsive pedagogy that helps us understand how to connect a sociocultural perspective of learning to classroom teaching" (Hall, 2001, p. 51). Within the multiliteracies framework, texts become the means through which students develop their literacy, and in doing so, gain access to products, practices, and perspectives of the target language and cultures they are learning about. A focus on literacies in "FL curriculum, instruction, and assessment creates a way to link the typically bifurcated parts of the undergraduate program [by bringing] textual analysis to introductory and intermediate courses, as well as language development to advanced courses" (Paesani et al., 2016, p. 14) (Table 5.2).

Table 5.2 Students' self-assessment: mean of all linguistic and cultural items from course surveys

Unit	Goals	SD/D	NA/ND	SA/A
1 The clothing industry: Sustainable	Linguistic	1.1%	6.1%	92.8%
consumption and production	Cultural	4.4%	4.4%	91%
2 Food as an intangible cultural heritage	Linguistic	0%	4%	95.9%
	Cultural	1.3%	13%	85.7%
3 Sustainable cities and communities	Linguistic	4.5%	9%	86.3%
	Cultural	0%	4.5%	95.4%
4 Migrants and the world they shape	Linguistic	2.6%	8.6%	88.7%
	Cultural	1.5%	4.5%	94%

Note: Abbreviations for tables: Strongly disagree (SD), Disagree (D), Neither agree nor disagree (NA/ND), Strongly agree (SA), Agree (A).

Curriculum Development

Once the content and pedagogical frameworks were defined, the curriculum development process followed a backward design model (Wiggins & McTighe, 2005), a three-stage approach whereby curricula must be built 'backward' by: (a) identifying the goals at the program level (i.e., what students should learn); (b) determining acceptable evidence that goals were achieved (i.e., how to assess that students have learned what they were supposed to); and (c) developing materials and learning experiences and instruction (i.e., specific instructional materials to help students reach the goals). Adhering to this model, the first step was to identify the goals of the course. For each unit, literacy, linguistic, cultural, and sustainability-oriented goals were specified. Literacy goals (adapted from Paesani, 2017) and sustainability goals, can be seen in Tables 5.3 and 5.4, respectively, in the Course evaluation section of this chapter.

The next step was to determine what evidence would be acceptable to confirm the desired content was learned, and the literacy, linguistic, cultural, and sustainability competencies developed. For this purpose, a set of project-based tasks was designed for each unit, an approach to language learning and assessment increasingly advocated for in L2 education (see Ferry, this volume; Cattell & Kleinhans, this volume). These projects required that students design texts of different modalities (e.g. written, oral, audiovisual, virtual), allowing instructors to assess students' progress continuously and thoroughly. Finally, the learning experiences and specific tasks used in each class session were designed. These tasks carefully scaffolded language and content, helping students successfully attain (and show evidence that they had attained) the desired goals.

Because different instructors teach this new course, a common metalanguage to communicate with them was needed. To this end, the modes of communication (i.e., interpersonal, presentational, and interpretive) proposed in the World-readiness Standards for Learning Languages (National Standards Collaborative Board, 2015) were adopted and adapted, creating language- and content-oriented can-do statements for each one of the thematic units, as well as for each of the class sessions contained in them. These can-do statements

Table 5.3 Students' self-assessment: mean of all literacy-oriented items from course surveys

Can-do statements: I can...	SD/D	NA/ND	SA/A
• Create and use culturally appropriate language in oral and written contexts	1.5%	3%	95.5%
• Understand and interpret culturally appropriate language in oral and written contexts	0%	13.6	86.3%
• Use a range of language forms appropriately and accurately	7.5%	9%	83.4%
• Understand cultural products, practices, and perspectives of the Spanish-speaking world and relate them to my culture	0%	1.5%	98.5%
• Analyze and interpret cultural content of oral, written, and visual texts of various genres	0%	9%	90.9%

Table 5.4 Students' self-assessment: mean of all sustainability-oriented goals from course surveys

Can-do statements: I can...	SD/D	NA/ND	SA/A
Unit 1	0%	4.4%	95.5%
• identify how the production of clothing items may have a negative impact on the environment.	0%	1.3%	98.7%
• identify how your own consumption habits may have a negative impact on the environment.	1.5%	7.7%	90.8%
• identify how your own consumption habits may be linked to human rights issues related to the production of clothing items (Latin America).			
Unit 2	7.8%	8.2%	84%
• explain why certain traditional foods in the Spanish-speaking world are important for transmitting cultural heritage (according to speakers from pre-assigned countries).			
Unit 3	3%	13.6%	83.3%
• identify the main sustainability challenges that residents of mainly Spanish-speaking communities face.	1.5%	1.5%	97%
• identify differences and similarities regarding sustainability amongst cities in the Spanish-speaking world.			
Unit 4	1.5%	7.5%	90.9%
• identify some of the challenges of migrants regarding access to jobs, quality education.			

were used to communicate each unit's specific goals to both students and instructors teaching the various sections of the course, and they were later used for program evaluation purposes. To coordinate efforts amongst instructors and make sure that sections effectuated the goals in the can-do statements, one of the authors of this chapter held weekly meetings with all instructors to discuss the rationale and sequencing of all synchronous and asynchronous tasks, select and review the texts for the upcoming class sessions, and brainstorm best pedagogical strategies. Thanks to this community of practice, each section steered the class in the same direction, ensuring the goals were effectively met. In the following subsections, we offer a detailed description of the four units comprising the new course (see Table 5.1). Each subsection explains how sustainability issues were presented and how linguistic and thematic content was articulated and scaffolded to properly serve the beginning Spanish learners taking this course.

Unit 1. The Clothing Industry: Responsible Consumption and Production

Clothing items are a subject included in most, if not all, elementary language courses. Clothing, and particularly the fashion industry, is also relevant as it pertains to some of the sustainability goals pursued by the UN (e.g. responsible consumption and production, decent work, clean water), which is why they have assembled globally coordinated initiatives such as the UN Alliance for Sustainable Fashion. However, traditional teaching materials limit their scope to describing clothing items and expressing likes and dislikes, missing an invaluable opportunity to delve into compelling sustainability issues. There are ways to go beyond this traditional approach by complementing or creating materials that allow for in-depth critical analyses of rather intricate issues, so long as content and language are carefully scaffolded. In the unit developed to tackle these issues, a somewhat traditional path is initially followed: First, students discuss their fashion preferences, describing what they wear daily, how clothes and colors function as cultural symbols in their countries, in politics, in advertising, etc. These introductory discussions serve to familiarize students with basic lexico-grammatical tools (e.g. clothing vocabulary and frequent verbs in the present tense, agreement) and a range of textual genres (e.g. advertisements, opinion pieces, tweets) that will be later deployed, as the unit progresses, to execute thought-provoking and content-rich tasks.

For instance, to articulate more substantive critical analyses, students examine their consumption habits and the impact that these habits may have on the environment. They do this by learning how much water is needed to produce different clothing items, and then tabulating how many liters of water it takes to manufacture the clothes they wear weekly. Second, students study the differences between fast fashion and slow fashion and their impact on the environment. These sessions dedicated to fast/slow fashion also seek to raise awareness of social sustainability issues related to decent work and human/labor rights in the Spanish-speaking world and elsewhere. As a case in point, students are asked to go through their closets and gather information about the places where their clothing items were manufactured (e.g.

Guatemala, Nicaragua), a task followed up with independent research on what the working conditions are in these countries (in English, to be submitted as homework). This topic leads naturally to a class discussion, in Spanish, on feasible strategies everyone can execute to reduce their negative footprint on the environment and human rights violations (e.g. repurposing our cloths, avoiding certain fabrics or brands, checking the label when we buy a product), a thorny but important conversation that often goes unchecked in language courses. As a final hands-on project for this unit, students are asked to develop a real, environmentally responsible clothing item using any sustainable materials at their disposal. Many students, for instance, have used plastic bags and repurposed them to create purses, bracelets, or worn-out shirts to make reusable bags; other students have undone old or unfashionable sweaters to knit new fashionable items. This last project, whose outcome is a tangible, sustainable product that students present and describe in class, inspires students and instructors to reflect upon creative solutions to shared environmental and human rights challenges that, to a certain extent, are everyone's responsibility.

Unit 2. Food as an Intangible Cultural Heritage

Local foods and ingredients represent a precious cultural manifestation. Along with oral traditions, festive events, or traditional crafts, food is recognized by the United Nations Educational, Scientific, and Cultural Organization (UNESCO) as intangible cultural heritage. As this organization notes, understanding a community's intangible cultural heritage helps "with intercultural dialogue, and encourages mutual respect for other ways of life" (UNESCO, n.d.). Aware of the importance of traditional foods and ingredients as cultural heritage, this unit explores various Hispanic cuisines and ingredients to try to fully understand their composition, history, cultural value, and whether they can be found in the Greater Boston area where the university is located or in students' places of origin. This latter point is crucial to understanding that access to traditional foods and ingredients represents one way to safeguard people's cultures, especially in diasporic communities, and therefore constitutes a matter of cultural sustainability. To introduce students to basic food vocabulary, the first weeks of this unit are dedicated to discussing different types of diets and the foods that students like or dislike, and how they can potentially contribute to their health and wellness. During this introduction, students also describe and analyze how in their own communities or countries, food is associated with specific cultural events (e.g. weddings, birthdays) or traditions. To segue into a discussion of food in the Spanish-speaking world, the class zeros in on one ingredient, corn, and studies its relevance in Mesoamerican cultures, framing the conversation within three thematic dimensions: Culinary, mythical, and industrial.

To increase students' exposure and understanding of other communities and cultures, the course incorporates TalkAbroad. This online platform offers face-to-face interactions with native Spanish speakers from almost every country in the Spanish-speaking world. Through TalkAbroad, students complete a 15-minute interview and obtain information about traditional food(s) in different countries. This task requires that students become knowledgeable on the 'foreign' cuisine of their TalkAbroad partner's country and determine the cultural significance of different local recipes. They need to find out when specific local dishes are served, their history, the meaning of their names, and why they are essential for the target culture(s). Once students have completed this task, they must cook a traditional dish from this country (or eat it at a local restaurant) to know what it tastes like. To demonstrate their expertise, at the end of this unit, students must use the linguistic tools and content they have learned to describe a country's cuisine to the class and to argue for this cuisine to be considered an intangible cultural heritage of humanity.

Unit 3. Sustainable Cities and Communities

According to data from the UN, despite occupying only 3% of the Earth's land, cities contain more than half of the world's population, consume up to 80% of the world's energy, and generate 75% of carbon emissions (UN, n.d.). Therefore, cities are fertile ground to explore almost all social, cultural, and environmental sustainability issues contemplated in the UN's 2030 goals for sustainable development. In this third unit, the focus is mainly on social issues, such as access to basic services (e.g. efficient transportation systems, access to healthcare, high-quality education), inequality (gender- and race-based, income-based), and economic opportunity (e.g. decent and stable employment). The unit opens with very structured and scaffolded analyses of authentic, up-to-date interactive data from sustainability rankings made by recognized organizations (e.g. IESE Business School, Arcadis). These data provide useful vocabulary and a sociological and statistical framework on which to base class discussions. Through these texts, students are exposed to the dimensions and specific parameters considered in determining what makes one city more sustainable than others. Importantly, these indexes are easy to navigate, even for beginning L2 learners, because they mainly provide isolated words, often cognates (e.g. categories such as 'education, transportation') and numbers. Linguistically, these categories and numbers are exploited to practice comparative structures while making meaningful comparisons between cities in Latin America, the US, and Spain that we further analyze in the following weeks.

Once students have learned basic vocabulary to speak about sustainable cities and communities, they spend the following few classes analyzing in more detail the most sustainable city in Spain (i.e., Madrid) and Latin America (i.e., Santiago, Chile). For Madrid, for example, they carry out a multifaceted exploration of the city's neighborhoods through a wide variety of perspectives and multimodal and multimedia texts that, albeit linguistically simple, are semiotically charged (e.g. street art, posters, Google virtual tours, Twitter feeds, pictures). They zoom in on Madrid's most multicultural neighborhood, Lavapiés, and discuss how touristification has triggered a gentrifying transformation that might be changing downtown indefinitely. Despite being a complex topic to tackle, discussions are made linguistically manageable without sacrificing content by progressively building up linguistic and thematic complexity. First, students develop the ability to describe the neighborhood in general terms, commenting on what buildings look like, what street art seems to suggest about its residents, and so on. Then, students examine simple yet authentic texts (e.g. maps, tourist guides) and complete a virtual tour to see for themselves how high-end stores and restaurants geared towards tourists are mushrooming in places that served as meeting spots for residents. Finally, the reading and analysis of a series of tweets make it possible for students to learn, in the residents' own words, how rents are too high and how stores have adapted to satisfy tourists' needs rather than the neighborhood's traditional population.

The final task for this unit looks to analyze further some of the most sustainable cities in the Spanish-speaking world. To this end, students use Talk-Abroad to complete a 15-minute conversation with a native speaker and obtain that person's perspective on their own city. The conversation is relatively easy for students to conduct because they come to it with prepared questions about different dimensions of sustainability.

Unit 4. Migrants and the World they Shape

Given the local nature of the Spanish language and the various Spanish-speaking communities in the US, one of the goals of the new curriculum is to develop an understanding of Latinx communities' linguistic practices, cultures, and position within US society as it pertains to living standards, access to services, and immigration policies, all of which are issues related to sustainability. As some scholars have noted (e.g. Alonso, 2006; Alvarez, 2013), Spanish programs in the US should de-foreignize their curricula, at least to some extent, and include content about local Spanish-speaking communities. Their argument is not unfounded: in a recent survey among Spanish language educators, participants indicated that they were more concerned about providing a balanced representation of international Spanish-speaking cultures than about how US Spanish-speaking communities were depicted, or their issues addressed, in current materials or courses (Brown & Thompson, 2018).

To examine identity and migration in local Latinx communities, the first step in Unit 4 is to untangle the concepts of 'Latino/a/@/x/e' and 'Hispanic.' In scrutinizing these concepts, class materials and discussions highlight the diversity of US Latinos/as/@s/xs/es (e.g. ethnoracial, gender, political, socioeconomic, linguistic). To illustrate this diversity, students conduct independent research and present to the class the life stories of a very diverse group of Latinos/as/@/xs/es, many of whom were born and raised in the US. At this point, students dissect demographic data to, among other things, demystify the widely held notion that Latinx communities are primarily migrant communities. This review leads naturally to a discussion of diasporic Latinx communities in the country: Who they are, where they live, what they do, etc. The subsequent class sessions are exclusively dedicated to one group of migrants: The so-called 'dreamers.' To portray the economic and educational inequalities and hardships that result from being undocumented in the US context, students analyze audiovisual and written narratives of recent child migrants.

The scope of this unit is expanded by including migration within the Spanish-speaking world. To this end, and after analyzing migration data in class, students interview native speakers from Venezuela, Colombia, Spain, Peru, and Costa Rica. This group of countries offers contrasting situations, some currently undergoing mass emigration due to political crises, others instead welcoming tens of thousands of migrants due to economic and social prosperity. In their TalkAbroad interviews, which must be reported in this unit's project-based assessment, students focus on the origin of migrants in their TalkAbroad partner's country, the type of jobs they find, the perception nationals have of them, etc. In the final part of the interview, students ask their partners if they would consider migrating to their countries of origin had they been born elsewhere. This last and difficult question is one that students themselves reflect upon in class on the last day of the semester.

Course Evaluation

A 'use-based' evaluation (Davis & McKay, 2018; Norris, 2016) was conducted to better understand and assess the effectiveness of the new course in achieving its literacy, linguistic, cultural, and sustainability-oriented goals. This evaluation intends to go beyond mere judgments of program value. Instead, it leads to useful curricular adjustments and changes (Davis & McKay, 2018) by the intended users, in this case, the two language faculty members overseeing the project. For this evaluation, both students ($n = 72$) and instructors ($n = 5$) completed anonymous surveys via Qualtrics. A follow-up 30 to 60 minute interview was conducted with each instructor to tap into their thoughts and ideas. These interviews were particularly valuable to gauge further the effectiveness and usefulness of all the different course components.

Students' surveys were administered at mid-semester and the term's end. In the mid-semester survey, students were asked to self-assess the degree to which they thought they had achieved the linguistic, cultural, and sustainability-oriented goals of Units 1 and 2. The survey administered at the term's end was similar, except that it inquired about the goals set for Units 3 and 4. For the last survey, an additional set of goals was adapted from Paesani (2017), measuring students' perception of their literacy development throughout the semester. In these surveys, students were presented with a matrix of can-do statements that represented the main goals of the units and the course overall (in the case of literacy goals). Using a Likert scale, students indicated their degree of agreement with each item. These matrix tables were followed by a series of open-ended questions that allowed students to expand upon their self-assessments.

The results from the surveys indicate that most students thought they had achieved the expected learning outcomes, an assessment that instructors shared. Between 86% and 95% of students agreed that they had met the targeted linguistic and cultural goals (see Table 5.2) and between 83% and 98% agreed they had met the literacy goals (see Table 5.3). The items focusing exclusively on sustainability-oriented goals, used to operationalize sustainability literacy, also received high scores in all units (see Table 5.4). These high scores were particularly significant for this project because of its attempt to create an educational experience that transcended the confines of the classroom. This enhanced sustainability literacy in all dimensions (social, cultural, and environmental) represents a tangible contribution towards nurturing critical locally and globally aware dispositions in the students taking this class.

In general terms, these results demonstrate that students found the course effective at advancing all its goals. The similarly high level of agreement that students reported for the linguistic, cultural, and sustainability goals suggests the course successfully balanced language and content instruction. As the comment below illustrates, students were particularly appreciative of the content-based nature of the class:

> Embedding linguistic concepts within the content-based units has been very helpful for learning and retaining new vocabulary and new grammatical structures. In my opinion, this style of learning has been more effective than other modes of language-learning pedagogy I have experienced, such as rote memorization. I also think it is very clever to combine the typical first-year vocabulary (clothes, food, etc.) with more interesting content issues such as sustainability, culture, migration, etc. This makes otherwise very basic content fresh and exciting.

Many students reported having increased their knowledge and critical awareness of sustainability challenges in Latin America and Spain. In describing their experience in class, they expressed gratitude for being exposed to unfiltered, first-hand accounts of what life is like in the Spanish-speaking world through authentic texts and conversations via TalkAbroad. These texts and exchanges allowed them to learn about and reflect on other cultures as well as their own. In the words of one student, these exchanges gave them a "better understanding of the world we live in, the problems we all face, and the varying perspectives of others."

The survey and interviews completed by instructors also revealed strong support for the new course. They indicated that students were more engaged with the class than in previous semesters, perhaps because they were "building skills as they also buil[t] interest and knowledge in the cultural context of the language they are becoming speakers of." This enhanced student engagement translated into tangible gains in terms of language development, sustainability literacy, and critical awareness, which were palpable in the quality of the work they turned in throughout the term. Instructors also appreciated the direct contact with native speakers:

> One of the lessons I took away from the recorded TalkAbroad conversations [in this class] was that students have a stronger ability to negotiate communication in the target language with an L1[native speaker] speaker than I expected. It was a rewarding experience both for the learners and for me.

Instructors also identified challenges that CBI entails, the most salient of which was the occasional pushback from some students accustomed to grammar-focused approaches and the difficulty in adequately scaffolding intricate content to beginning learners. The instructors, however, did not present these challenges in a negative light but rather as a 'worthwhile' experience that made them more aware of the need from instructors and students to have a clear understanding of the goals of the class, as well as the strategies being deployed to help them achieve those goals. As one of the instructors said,

> [It] is always said that a balance between cultural content and grammatical content is possible in the first levels of second language education. However, it had always been difficult for me to imagine, beyond a specific exercise, the application of this idea in systematic terms of an entire level. But I think this course is exemplary in terms of the possibility of establishing such a balance.

The follow-up interviews with instructors helped gather further information about students' achievement of the general goals of the class. During these interviews, there were careful and in-depth discussions about goals (in the students' or instructors' surveys) that did not reach at least 90% agreement. These collaborative discussions helped identify aspects that needed to be fine-tuned before the next iteration of the course. Ways to improve or modify specific materials and pedagogical approaches were outlined. For instance, some of the texts in Units 1, 3, and 4 were changed, and the sequencing of material was slightly modified in Units 2 and 3. As a case in point, a documentary about the environmental consequences of a hydroelectric project being developed near Santiago, Chile, was removed from Unit 3 because it elicited discussions that students were not linguistically prepared to have. To replace it, a series of newspaper headlines and cartoons about the same hydroelectric project was selected. More changes will indeed have to be considered in future iterations of this course.

Final Remarks: Lessons Learned

This chapter provided a detailed description of the transformation of a second-semester beginner Spanish language course at a northeastern university.

Students' positive evaluations of the first iteration of this course point to the benefits and strengths of building curricula around compelling sustainability-based topics related to local, national, and global Spanish-speaking communities. The students' survey results also indicate the advantages of using a CBI approach in conjunction with the multiliteracies framework to design courses and instructional experiences that help advance learners' content knowledge and their linguistic, cultural, sustainability, and literacy-based competencies. The feeling that they had learned "more than just Spanish" was mentioned repeatedly. For instance, one student said they appreciated the "multiple successful accounts to integrate grammar and broader language lessons into the fray of the discussion of social events and phenomena, which have mutually reinforced a deeper understanding of the language."

This experience was also rewarding for the faculty in charge of curriculum development and the instructors. The weekly meetings held throughout the semester were particularly valuable in creating an effective course in its initial implementation, helping instructors to fully understand the purpose of all course components, and allowing them to adjust texts, tasks, and assessments in real-time. In these weekly meetings, and later in the interviews that formed part of the course evaluation, instructors also reflected upon the lessons learned and the benefits of thinking about curricula in a 'backwards design' fashion, having a clear idea of all the goals and assessments before deciding how to approach individual class sessions and what texts to include. Instructors regularly highlighted the importance of providing "enough scaffolding activities as to ensure no one is discouraged from approaching the content in the target language," especially when the content was culturally and linguistically complex.

Curricular projects such as the one described in this chapter aim to contribute to current efforts at other higher education institutions to make foreign language literacy, even at beginning levels of instruction, a significant and integral part of students' liberal arts education. With courses like this, students can become more critically conscious of the challenges they share with Spanish-speaking communities, including those in the US, and how a better understanding of those communities' worldviews can help create a socially and environmentally responsible glocal citizenry for the 21st century and beyond.

Notes

1 For more information about Georgetown's SFS Spanish program's new curriculum, see http://www.aausc.org/AAUSCblog/9418161
2 Many successful collegiate curricular reforms in the US have used literacy-based approaches (e.g. Byrnes et al., 2010; Maxim et al., 2013; Paesani, 2017).
3 This endeavor is part of a larger curricular project that transformed the two-semester beginning Spanish sequence. The first-semester course focuses on social and cultural sustainability issues (e.g. sociopolitical representation, Latinx identity).

References

Alvarez, S. (2013). Evaluating the role of the Spanish department in the education of US Latino students: Un testimonio. *Journal of Latinos and Education, 12,* 131–151. https://doi.org/10.1080/15348431.2012.745405

Alonso, C. (2006). Spanish: the national foreign language, *ADFL Bulletin,* 37(2–3), 15–20.

Bernhardt, E., & Berman, R. (1999). From German 1 to German Studies 001: A Chronicle of Curricular Reform. *Die Unterrichtspraxis / Teaching German, 32*(1), 22–31. https://doi.org/10.2307/3531869

Bori, P. (2018). *Language Textbooks in the Era of Neoliberalism.* Milton: Routledge.

Brown, A., & Thompson, G. (2018). *The changing landscape of Spanish language curricula: Designing higher education programs for diverse students.* Washington, DC: Georgetown University Press.

Byrnes, H., Maxim, H., & Norris, J. (2010). Realizing advanced foreign language writing and development in collegiate education: Curricular design, pedagogy, assessment. *Modern Language Journal, 94*(s1), 1–235. http://www.jstor.org/stable/40985261

Cammarata, L. (2016). *Content-based foreign language teaching curriculum and pedagogy for developing advanced thinking and literacy skills.* New York, NY: Routledge.

Cammarata, L., Tedick, D., & Osborn, T. (2016). Content-based instruction and curricular reforms: Issues and goals. In L. Cammarata (Ed.), *Content-based foreign language teaching curriculum and pedagogy for developing advanced thinking and literacy skills.* New York, NY: Routledge.

Cope, W., & Kalantzis, M. (2009). "Multiliteracies": New literacies, new learning. *Pedagogies, 4*(3), 164–195. https://doi.org/10.1080/15544800903076044

Cubillos, J. (2014). Spanish textbooks in the US: Enduring traditions and emerging trends. *Journal of Spanish Language Teaching, 1*(2), 205–225. https://doi.org/10.1080/23247797.2014.970363

Davis, J.M., & McKay, T.H. (2018). *A guide to useful evaluation of language programs.* Washington: Georgetown University Press.

de la Fuente, M. J. (2006). Classroom vocabulary acquisition: Investigating the role of pedagogical tasks and form-focused instruction. *Language Teaching Research, 10*(3), 263–295. https://doi.org/10.1191/1362168806lr196oa

Dubreil, S., & Thorne, S. (2019). *Engaging the world: Social pedagogies and language learning* (Issues in language program direction 2017). Boston, MA: Cengage.

Hall, J. K. (2001). *Methods for teaching foreign languages: Creating a community of learners in the classroom.* Upper Saddle River, NJ: Prentice Hall.

Herman, D. (2007). It's a small world after all: From stereotypes to invented worlds in secondary school Spanish textbooks. *Critical Inquiry in Language Studies, 4*(2–3), 117–150. https://doi.org/10.1080/15427580701389417

Kern, R. (2000). *Literacy and language teaching.* Oxford: Oxford University Press.

Klimanova, L., & Hellmich, E. A. (2020). Putting local on the MAP: A model for engaging foreign language students with local cultures. *Foreign Language Annals, 54*(1), 158–184. https://doi.org/10.1111/flan.12493

Kramsch, C., & Vinall, K. (2015). The cultural politics of language textbooks in the era of globalization. In X. L. Curdt-Christiansen & C. Weninger

(Eds.), *Language, ideology and education: The politics of textbooks in language education* (pp. 11–28). Routledge.

Lyster, R. (2007). *Learning and teaching languages through content: A counterbalanced approach.* Amsterdam: Benjamins.

Lyster, R. (2017). Preface. In J. Valcke & R. Wilkinson (Eds.), *Integrating content and language in higher education: Perspectives on professional practice* (pp. 7–14). Frankfurt: Peter Lang.

Maxim, H. H., Höyng, P., Lancaster, M., Schaumann, C., & Aue, M. (2013). Overcoming curricular bifurcation: a departmental approach to curriculum reform. *Die Unterrichtpraxis, 46*(1), 1–26.

Melin, C. (2014). Program sustainability through interdisciplinary networking: On connecting foreign language programs with sustainability studies and other fields. In J. Swaffar & P. Urlaub (Eds.), *Transforming postsecondary foreign language teaching in the United States.* Dordrecht: Springer Netherlands.

Melin, C. (2019). *Foreign language teaching and the environment: Theory, curricula, institutional structures.* New York: The Modern Language Association of America.

Mills, N. A., Courtney, M., Dede, C., Dressen, A., & Gant, R. (2020). Culture and vision in virtual reality narratives. *Foreign Language Annals, 53*(4), 733–760.

MLA Ad Hoc Committee on Foreign Languages. (2017). Foreign languages and higher education: New structures for a changed world. *Profession, 2007*(1), 224–245. https://www.jstor.org/stable/25595871

National Standards Collaborative Board (2015). *World-readiness standards for learning languages* (4th ed.). Alexandria, VA: Author.

New London Group. (1996). A pedagogy of multiliteracies: Designing social futures. *Harvard Educational Review, 66*(1), 60–92. https://doi.org/10.17763/haer.66.1.17370n67v22j160u

Norris, J. (2016). Language Program Evaluation. *Modern Language Journal, 100*, 169–89. http://www.jstor.org/stable/44135002

Paesani, K. (2017). Redesigning an introductory language program: A backward design approach. *L2 Journal, 9*(1), 1–20. https://doi.org/10.5070/L29130408

Paesani, K., & Allen, H. W. (2012). Beyond the language content divide: Research on advanced language instruction at the postsecondary level. *Foreign Language Annals, 45*(S1), 54–75. https://doi-org.ezp-prod1.hul.harvard.edu/10.1111/j.1944-9720.2012.01179.x

Paesani, K., Allen, H. W., & Dupuy, B. (2016). *A multiliteracies framework for collegiate foreign language teaching.* Boston, MA: Pearson.

Parra, M. L. (2013). Expanding language and cultural competence in advanced heritage- and foreign language learners through community engagement and work with the arts. *Heritage Language Journal, 10*(2), 253–280. https://doi.org/10.46538/hlj.10.2.7

Patel, F., & Lynch, H. (2013). Glocalization as an alternative to internationalization in higher education: Embedding positive glocal learning perspectives. *International Journal of Teaching and Learning in Higher Education, 25*(2): 223–230.

Sato, S., Hasegawa, A., Kumagai, Y., & Kamiyoshi, U. (2017). Content-based instruction (CBI) for the social future: A recommendation for critical content-based language instruction (CCBI). *L2 Journal, 9*(3), 50–69. https://doi.org/10.5070/L29334164

Snow, M., & Brinton, D. (2017). *The content-based classroom* (2nd ed.). Ann Arbor: University of Michigan Press.

Swaffar, J., & Arens, K. (2005). *Remapping the foreign language curriculum: An approach through multiple literacies.* New York: The Modern Language Association of America.

UNESCO. (n.d.). *What is intangible cultural heritage?* UNESCO. https://ich.unesco.org/en/what-is-intangible-heritage-00003

Wiggins, G., & McTighe, J. (2005). *Understanding by design* (2nd ed.). Alexandria, VA: Association for Supervision and Curriculum Development.

6 Beyond the Language Requirement

Implementing Sustainability-Based FL Education in the Spanish Foundations Program

Raychel Vasseur and Yerko Sepúlveda

Introduction

The last ten years have seen an urgent call from different organizations, activists, and scholars worldwide to include education for sustainable development across disciplines. In 2015, the World Education Forum[1] (WEF) sought agreement from the international education community to promote inclusive and equitable education with sustainable development goals (SDGs). An outcome of the forum was the Incheon Declaration and Framework for Action declaring, among other things, that by 2030, "all learners acquire knowledge and skills needed to promote sustainable development, including [...] human rights, gender equality, promotion of a culture of peace and non-violence, global citizenship and appreciation of cultural diversity and culture's contribution to sustainable development" (UNESCO, 2016, p. 48).

It seems, however, that the environmental aspects of sustainability receive the most attention, especially in foreign language (FL) education (e.g., see Melin, 2019), while sociocultural aspects—including the critical function of language, communication, and culture in the achievement of sustainable development—have been somewhat neglected (Zygmunt, 2016). To recognize the need for more understanding of this area in sustainability education as a global focus (ter Horst & Pearce, 2010), as well as to create more intellectually stimulating FL learning experiences (Cammarata, 2016), the Department of Classical & Modern Languages & Literatures (CMLL) at Texas Tech University (TTU) developed and implemented a pedagogical initiative utilizing the 'Re-imagining Migration' framework, a collaboration between the Harvard Graduate School of Education's Project Zero and UCLA's School of Education and Information Sciences (Project Zero, n.d.a), in the third and fourth-semester Spanish program. The overarching goals were: (1) To promote sustainable FL education focused on quality education; (2) increase student linguistic and cultural competence, thus contributing to advancing departmental goals and promoting the necessary sensitivity to live in a globalized world; and (3) to intellectually challenge students.

This chapter will describe the curricular work implemented in a fourth-semester, general education requirement, Spanish course: This was a

DOI: 10.4324/9781003080183-9

project-based pedagogical unit (see Chapter 3, this volume) in which students discussed and reflected on the topic of migration throughout the semester, culminating with the creation of an artifact and project presentations. The migration project served as a catalyst to examine cultural understanding that contributes to the achievement of quality education as outlined in SDG 4.7, which states:

> By 2030, ensure that all learners acquire the knowledge and skills needed to promote sustainable development, including, among others, through education for sustainable development and sustainable lifestyles, human rights, gender equality, promotion of a culture of peace and non-violence, global citizenship and appreciation of cultural diversity and of culture's contribution to sustainable development.
>
> (UNESCO, 2016, p. 21)

First, we briefly discuss the theoretical underpinnings that served as a framework for the initiative. Then, we describe the context and the project design, followed by an evaluation of the project, including examples of student responses and products. Finally, we identify some challenges we found while implementing these changes and offer suggestions for overcoming said challenges.

Theoretical Underpinnings and the Call to Action

Sustainability-Based Education

The WEF framework for action invites stakeholders to rethink and redesign their programs to help students acquire the knowledge and skills to shape a sustainable future. However, as previously mentioned, the sociocultural perspective of sustainability education seems to be receiving less attention. However, from a culturally and socially responsive lens, "attempts to reform and design education programmes must focus on basic human rights—linguistic rights and social inclusion" (Babaci-Wilhite et al., 2012, p. 641). Likewise, at its core, SDG 4 refers to "rights-based and inspired by a humanistic vision of education and development, based on the principles of human rights and dignity, social justice, peace, inclusion, and protection, as well as cultural, linguistic and ethnic diversity and shared responsibility and accountability" (UNESCO, 2016, p. 24). As suggested by Badjanova and Iliško (2015) "teachers should encourage learners to be active participants in situations of social change rather than to prepare them to passively adapt to the social environment" (p. 138). Therefore, when learning an FL, students must explore and understand problems of different natures that affect the living world (Zygmunt, 2016), form the attitudes and values necessary to build a sustainable future (Pevneva et al., 2017), and utilize the new opportunities to question embedded prejudices and assumed realities (Cammarata et al., 2016). Within the field of FL education, the combination of the Incheon Declaration and SDGs has allowed for additional inter- and multi-disciplinary perspectives and curricular transformations (de la Fuente, 2019), thus increasing the visibility of sustainable education.

Sustainability (Content)-Based Foreign Language Education

Integrating content and language learning objectives has ignited interest among language professionals and curriculum specialists globally (Stoller, 2004), with goals ranging from more effective learner engagement and motivation to better education for a just society (Cammarata, 2016). At its core, "a content-based instructional approach is one in which language proficiency is achieved by shifting the focus of the course from the learning of language per se to the learning of subject matter" (Leaver & Stryker, 1989, p. 270). In this manner, content-based instruction (CBI) is one pedagogical strategy to effectively cross disciplinary boundaries while teaching FL and content material. Furthermore, by transforming FL themes to be 'thinking-rich,' to include topics like sustainability, rather than 'thinking-light' as they are typically covered (Martel, 2016), the curriculum is also able to target three of the five World-Readiness Standards for Language Learning—*connections, cultures,* and *communities* by providing a context in which students can think critically about the issues, materials, and experiences they explore, what information may have been left out of the conversation, and why (Martel, 2016).

The themes and content chosen for a CBI curriculum are critically important and should be selected with students' age, cognitive maturity level, language proficiency, and interests in mind (Met, 1998). They should also be selected according to programmatic goals, which, as suggested by Cammarata et al. (2016), should be to

> better stimulate students' motivation to engage *willingly* and *wholeheartedly* in the study of a language within and beyond the school setting and...be supportive of the development of learners' capacity to engage in deeper forms of thinking that will lead to intellectual autonomy.
>
> (p. 9, emphasis original)

One topic that has been successful in CBI for L2 learning at secondary and post-secondary levels is sustainability (de la Fuente, 2019; Melin, 2019), as it allows for the integration of cultural issues, specific vocabulary knowledge, and social practices with critical thinking and problem-solving tasks (de la Fuente, 2019; Jodoin, 2019; ter Horst & Pearce, 2010). Furthermore, "sustainability content provides opportunities for much needed, complex language production to reach advanced levels of proficiency" (de la Fuente, 2019, p. 134).

The department of CMLL at TTU strives to incorporate global sociocultural thinking in the Spanish language requirement courses. In an effort to stimulate proficiency, promote cross-cultural and cross-disciplinary connections, and develop more globally-minded students, a content-based Migration Project was developed for a fourth-semester Spanish course. The following section describes its implementation.

The Migration Project

The Context

The Spanish Foundations program is based on communicative language teaching pedagogy; it utilizes a pedagogically appropriate textbook and a flipped methodology. Since the summer of 2019, both the third- and fourth-semester Spanish courses include content-based instructional units that are project-based. This project has included pedagogical training for course instructors[2] as well as minor modifications and updates. This chapter describes The Migration Project, the fourth-semester course, sustainability-based unit that impacts approximately 1,000 students per year. TTU is a Tier-1, Hispanic-serving research institution with over 40,000 students located in West Texas. As part of a graduation requirement, many students must complete through either the third or fourth semester of a FL. There is a separate course sequence for students who identify as Heritage language speakers.

Re-Imagining Migration

To fully embrace a strong focus on sustainability-based FL education, the lifelong learning opportunities specified in SDG 4 were foundational. These opportunities are not only framed within cognitive aspects of learning and life skills but also within broader contexts such as "rapidly changing labor markets, technological advances, urbanization, migration, political instability, environmental degradation, natural hazards and disasters, competition for natural resources, demographic challenges, increasing global unemployment, persistent poverty, widening inequality and expanding threats to peace and safety" (UNESCO, 2016, p. 26). When designing the content-based course projects (both for the third- and fourth-semester Spanish courses), we worked within target 4.7 (see *Introduction*, this chapter) of the Sustainable Development Goals, and contextualized it within migration, one of the broader contexts within which education must be seen (UNESCO, 2016).[3] We situated the project within the context of migration because studying it is part of the vision and strategic plan to achieve SDG 4 by 2030 (UNESCO, 2016); at the same time, critically examining the migration narrative was—and still is—extremely relevant to our population of students at TTU.

Due to the nature of the project, the Re-imagining Migration framework[4] (Project Zero Re-imagining Migration, n.d.a) was used as a resource to explore materials, inform pedagogical decisions, and analyze the resulting data. The Re-imagining Migration framework requires and is designed to create a shift in mindset surrounding migration by specifically presenting migration as an opportunity rather than a pressing challenge. It operationalizes migration as "a shared condition of our past, present, and future in order to develop the knowledge, empathy, and mindsets that sustain inclusive and welcoming communities" (Project Zero, Re-imagining Migration, n.d.a). While using it, educators and students have an opportunity "to re-frame migration…as an opportunity to re-imagine a new approach to education—one destined to benefit all" (Boix Mansilla, n.d.).

The Cycles

The Migration Project is conducted throughout one semester in four cycles. The goals are two-fold—to discuss and reflect on migration, as defined by Re-imagining Migration, and to develop the dispositions,[5] knowledge, skills, and attitudes to achieve SDG 4.7. During the project, students work with an array of migration-related content in different formats (e.g. newspaper articles, videos, and images) and engage in individual and collective reflection, culminating the project by creating a cultural artifact—an accordion book or a narrative essay. Each cycle, conducted every two to three weeks, comprises three steps: (1) Interact and engage with content; (2) submit an essay; and (3) work collaboratively during class to unpack content and document collective ideas with thinking routines. Steps 1 and 2 are conducted individually before class; step 3 is conducted during class as a collective group. Following the fourth cycle, students conduct an interview and create their final artifact (see Figure 6.1).

Step 1: Interact and Engage with Content

Students engage and reflect on the content of the material they are assigned before class. Students read, watch, or listen to specific sources as homework before engaging with the content in class with their peers. According to Willingham (2019), "we interpret new information in light of what we already

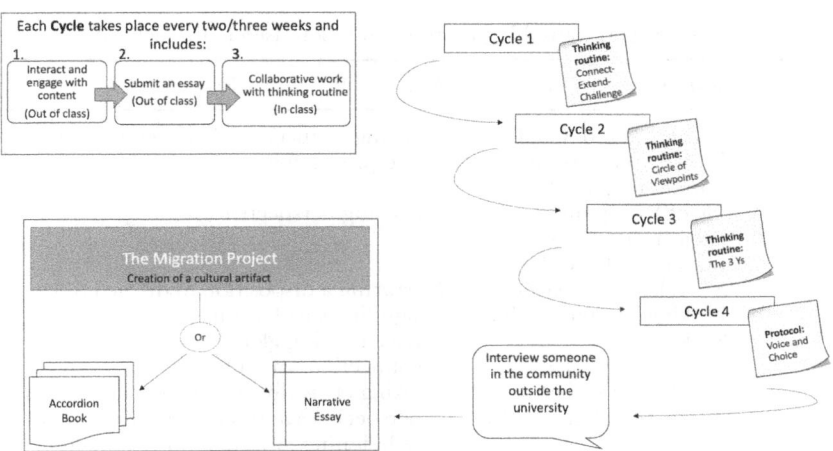

Figure 6.1 Project cycle of the migration project

know. The right preparation makes new learning easier" (2019, p. 12). Thus, this first step prepares them and activates prior knowledge. It helps students connect ideas with personal, lived experiences of their own or of those they know.

Step 2: Submit an Essay

After students have engaged with content outside of class, they submit a 250-word essay inviting them to summarize the main ideas of the content they explored and to respond to essay prompts that connect the materials with a larger migration frame and its impact on local and global patterns of people moving around the world. While students are encouraged to write in the L2, they are given the option to use English to help scaffold (Lantolf, 2000) their ability to process and reflect on complex subject matter for the first time. Engaging with the material thoroughly and documenting their ideas in their first language (L1) prepares them to engage in class L2 production tasks, some of which require superior-level skills, such as hypothesizing and defending and supporting opinions (ACTFL, 2012). They prepare for each of the five Re-imagining Migration class discussion sessions in a similar way.

Step 3: Work Collaboratively in Class

Once students have engaged with materials and written about their connections, they work collaboratively in class to further unpack the assigned content. They document their individual and collective reflections through *thinking routines* (see Table 6.1) (Project Zero's thinking routines toolbox, n.d.b), or simple patterns of thinking that can be used repeatedly and

Table 6.1 Project Zero's thinking routines and tools used in project cycles

Cycle	Thinking routine/protocol	Purpose
1	Connect, Extend, Challenge (Project Zero's thinking routines toolbox, 2019a)	Drawing connections between new ideas and prior knowledge
2	Circle of Viewpoints (Project Zero's thinking routines toolbox, 2019b)	Exploring perspectives
3	The 3 Whys (Project Zero's thinking routines toolbox, 2019c)	Nurturing a disposition to discern the significance of a situation, topic, or issue keeping global, local, and personal connections in mind
4	Protocol voice and choice (Agency by Design, 2018)	Looking critically at content, considering perspectives and representation, redesigning or reimagining that content from one's perspective.

easily folded into learning (Perkins, 2003). The goal of these discussions is to deepen their understanding of the topic and work with the vocabulary to discuss it in Spanish. They offer directive language and steps that trigger thinking while offering language chunks to document students' voices in the FL. A new thinking routine was introduced for each new topic and discussion.

For example, the Circle of Viewpoints Routine (see Figure 6.2) (Project Zero's thinking routines toolbox, 2019) invites learners to explore diverse perspectives. It offers three prompts to activate thinking plus an extension as a fourth step. The students engage with this routine by responding to (1) *Estoy pensando acerca de…desde el punto de vista de…* (I am thinking of [the topic]

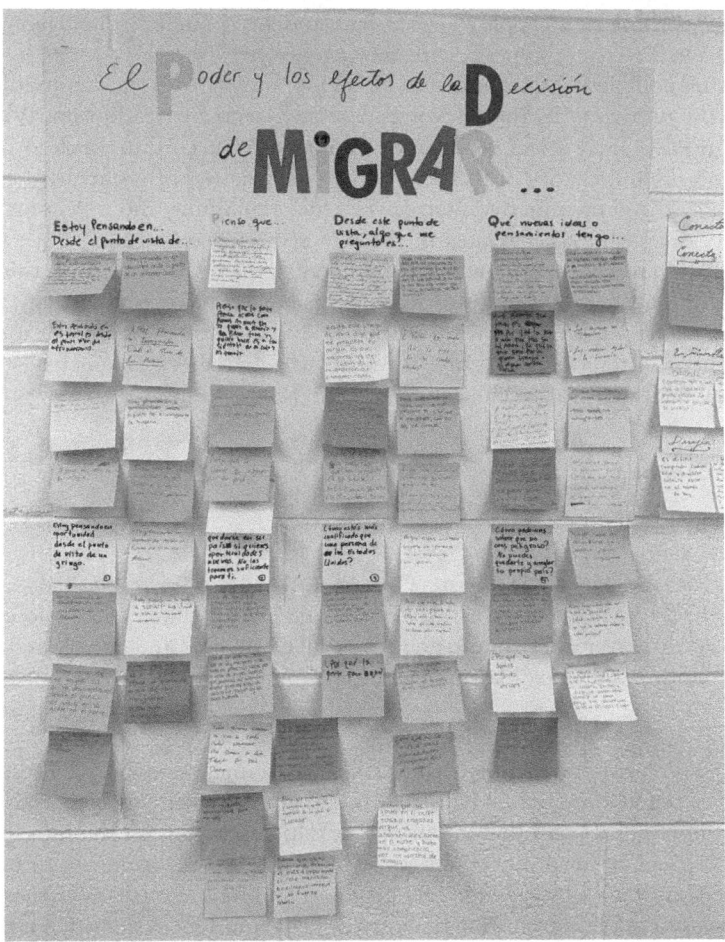

Figure 6.2 Circle of viewpoints in Spanish

from the point of view of [the chosen viewpoint]), (2) *Pienso que...* (I think... [describe the topic from the viewpoint]), (3) *Una pregunta que tengo desde este punto de vista es...* (A question I have from this viewpoint is...), and (4) *¿Qué nuevas ideas o pensamientos tengo?* (What new ideas or understandings about the topic do I have?).

Through thinking routines, instructors scaffold the development of thinking skills, allowing students to build a range of thinking strategies (Ritchhart & Church, 2020). When used in the FL classroom, thinking routines also prompt target language (TL) use and scaffold thinking in a developing linguistic system, allowing students to monitor and evaluate comprehension and production. For example, zooming in on Figure 6.2, responding to *Pienso que ...* (I think that...) a student responded *"pienso que las personas nos ven como las estudiantes malos, pobre, menos inteligente, matones. Creo que estos estereotipos son incorrectos."* (I think that people see us as bad students, poor, less intelligent, thugs. I believe these stereotypes are incorrect.)

After completing the four cycles, students connect with the community outside the university by interviewing someone[6] to create a cultural artifact (see Figure 6.1). They select between two options: An accordion book to recount the personal life story of the person they interviewed or a narrative essay to introduce and narrate the interviewee's life, explaining and reflecting on the interviewee's life as it relates to the topics discussed in class. Once the interviews have been conducted, the students create the cultural artifact and present it to their classmates. The students have three options—interview an immigrant, a US citizen with at least one foreign-born parent, or a grandparent or distant relative. Once the interviews have been conducted, students create their cultural artifacts.

The Accordion Book

An accordion book

> describes any book having a folded, rather than sewn, text block, where the pages are pleated – folded back and forth...[like]...the bellows of an accordion, rather than being folded like a map – and are viewed by expanding the book like an accordion.
>
> (Thomas & Thomas, 2002)

Students can create their accordion books with images, maps, pictures, direct quotes, reflections, and fabric pieces. Their comments are written in Spanish, but direct quotes can be written in English, depending on the interviewee's language. The accordion book must be original and demonstrate both an understanding of the phenomena explored and intercultural awareness. Students present their accordion book to the class and submit a written reflection about their process and what they learned (Figure 6.3 represents an exhibit with all accordion book projects).

Figure 6.3 Accordion book projects

Data Analysis

After the initial content analysis of the data the authors chose the five *core dispositions* from the Re-imagining Migration educational framework to categorize emerging themes in the data. While the framework's purpose is not to evaluate student learning outcomes, these dispositions can be conceptualized to design learning goals to prepare students for living in a more sustainable world. The five dispositions of this framework (see Table 6.2) are (1) understanding perspectives; (2) inquiring in a world shaped by migration; (3) communicating and building relationships across differences; (4) recognizing inequities; and (5) taking action toward inclusive and sustainable societies. They are designed to increase sensitivity toward mobility, diversity, and preparation for a more sustainable world on the move. The implementation of the semester-long course project showcases these five dispositions. The dispositions were used as the lens to view students' development towards the achievement of SDG 4.7, which has as its goal that all learners acquire the knowledge and skills to promote sustainable development through education and appreciation of cultural diversity (UNESCO, 2016). Furthermore, the Incheon Declaration specifies the need for education systems to be relevant and have the ability to respond to rapid changes around the world (UNESCO, 2016). Therefore, the five dispositions are a useable tool to document inclusive and equitable quality education. Although the five dispositions were not used to teach the content, they present an avenue to measure the achievement of some of the indicative strategies outlined in the Incheon Declaration (UNESCO, 2016) to help build peaceful and sustainable societies. For example, all five dispositions are represented in the second indicative strategy for SDG 4.7, which states,

> [t]he content of such education must be relevant, with a focus on both cognitive and non- cognitive aspects of learning. The knowledge, skills, values, and attitudes required by citizens to lead productive lives, make informed decisions and assume active roles locally and globally in facing and resolving global challenges can be acquired through education for sustainable development and global citizenship education, which includes

Table 6.2 Five Core Dispositions (Re-imagining Migration, n.d.) used for data analysis

Disposition	Subcategory
Understand perspectives (others and one's own)	– Understand and value one's perspective – Understand and empathize with others – Appreciate the dynamism of cultures and perspectives
Inquire (in a world shaped by migration)	– Feel connected and a sense of belonging to a larger human story – Ask relevant and informed questions – Gather, weigh, and reason with evidence – Managing complex ideas, contexts, and emotions
Communicate (and build relationships across differences	– Build bonding and bridging relationships – Appreciate respectful dialogue – Employ and interpret symbol systems and languages – Recognize and reflect on communication and relational challenges
Recognize inequities	– Recognize inequities and power disparities in daily experiences – Seek to understand historical and current inequities – Understand one's positions vis-à-vis inequities
Take action toward inclusive and sustainable societies	– Sense of belonging – Be sensitive toward opportunities – Employ understanding, voice, and capacity for influence – Employ a repertoire of civic engagement tools – Sense of self-efficacy as a change maker

peace and human rights education as well as intercultural education and education for international understanding.

(p. 49)

Table 6.2 shows the five core dispositions and their subcategories from the Re-imagining Migration Framework used to code the data.[7] The authors coded one of each of the four essays and one final course project together to calibrate their coding decisions, then they analyzed the students' final projects independently and finally came together to compare the results and discuss their analysis until an agreement was reached. All thinking routine images were coded together.

Results and Discussion: Five Dispositions for a World on the Move

Understanding Perspectives: Others and One's Own

The powerful words students wrote demonstrated an appreciation for a variety of perspectives, a disposition crucial for sustainable development due to the importance of perspective taking in "global citizenship and appreciation of cultural diversity and of culture's contribution to sustainable development" (UNESCO, 2016, p. 21). One student wrote in her final project reflection, "We automatically assume we know everything about a certain culture or country, but there are so many different stories that need to be told and listened to so we can fully understand" (Kim,[8,9] Essay 1). Students also showed evidence of challenging their positions and understanding of migration and others' perspectives. One student wrote in her final reflection,

> We all like to believe that we are open-[minded] and welcoming to other cultures. I always thought I had an open mind, but I have learned throughout studying and working on this course project that I am not as welcoming as I want to be.
>
> (Diana, Final Reflection)

Another student recognized other blind spots they and others might have, demonstrating his appreciation for cultural diversity and the importance of other culture's contributions, two core aspects of SDG 4.7, writing:

> The dangers of a single story could be the loss of culture and heritage because if you only know one thing, that's all you will see. This leads to ignorant people in the world not by choice but by not knowing better, if all you see on tv or read you'll have only one image and thought of people which can be highly inaccurate.
>
> (Jaxon, Essay 1)

Most importantly, though, especially during the current polarized political climate, students realized that considering multiple perspectives was critical. One student reflected, "*Cuando se tiene una sola historia, es fácil a tener ideas falsas sobre un grupo de personas*" [When you only have one story, it's easy to have false ideas about a group of people] (Dina, Essay 1). This theme continued to develop throughout the semester for Dina. In her final reflection, she wrote, "the ability to consider someone else's perspective is important in beginning to understand another person's version of 'normal.' Just because something is different doesn't make it bad" (Dina, Final Reflection). Another student also highlighted the multiple perspectives he had, reflecting:

> I had a wide variety of thoughts throughout this course project. I had a lot of preconceived notions about immigration and what goes into it and looking back I feel bad about those. This was a project that forced us, regardless of what we previously believed, to think deeper or from a different angle than we had ever done before. I really enjoyed how we took more than one perspective on immigration and migration and looked at various stories in our nation's immigration history beyond just what is currently happening right now. Even within exploring different stories of

migration we were pushed to think from a different perspective than we
might otherwise think.

(Andrew, Final Reflection)

As seen through the lens of the *understand perspectives* disposition (Re-imagining
Migration, n.d.), through reflections students show they had begun to acquire
the knowledge and skills needed to promote sustainable development through
education due to their ability to appreciate cultural diversity and to appreciate
other cultures' contributions to sustainable development and to society.

Inquire: In a World Shaped by Migration

The second disposition contains four subcategories related to a person's ability,
inclination, and sensitivity to: (1) see migration as a shared experience; (2) demonstrate curiosity by asking informed, relevant questions; (3) make sense of issues
related to migration by working with quality data; and (4) manage complex ideas
and emotions, and draw conclusions with compassion (Boix Mansilla, n.d.).

To demonstrate this disposition, students expressed feeling connected to
immigration as a larger human story for several reasons, including because
they and their families were immigrants. One student wrote, "*Mi voz está rep-
resentada porque vengo de una familia de immigrantes mexicanos* [My voice is represented because I'm from a family of Mexican immigrants]" (From Protocol
Voice & Choice, Cycle 4) after they reflected on an art installation and interviews with the artist, JR, as part of an in-class thinking routine.

Other students felt connected to this larger story of migration while simultaneously examining migration with compassion and handling complex ideas
about the situation, realizing the local stories they knew of migration:

> relate to global patterns of people moving around the world...People are
> constantly on the move. All for different reasons. Some fleeing persecu-
> tion or government oppression, some migrate just to begin a new story
> in a different place. Localized experience reflects the bigger picture...it's
> happening all of the world at all times. We can put this all to use in the
> larger frame by always keeping it at the forefront of our mind.
>
> (Payton, Essay 2)

Another student expressed a similar compassionate sentiment when considering the power and the effects of the decision to migrate while utilizing the
thinking routine, the 3 Whys and answering the question: *¿Qué nuevas ideas o
pensamientos tengo...* [What new ideas or thoughts do I have...] saying, "*Por que
los inmigrantes son vistos como malos? Todos somos los inmigrantes* [Why are immigrants seen as bad? We're all immigrants]" (Thinking Routine the 3 Whys,
Cycle 3). The lens of *inquire* promotes the development of learners' abilities to
gather knowledge and manage complex emotions and ideas they have about
sustainability, including topics like migration.

Communicate and Build Relationships across Difference

The third disposition describes someone who has the ability, inclination, and sensitivity to: (1) Build relationships both within and across communities through caring and respectful interactions; (2) appreciate and implement respectful, culturally sensitive dialogue; (3) use and interpret language and symbolic systems of communication to build relationships; and (4) recognize and reflect on challenges in communication (Boix Mansilla, n.d.). In addition to continuing to develop students' perspectives, this disposition can be related to promoting a culture of peace and non-violence as well as global citizenship for sustainable development.

The most obvious way students employed and interpreted symbols and languages were in the fourth cycle through the interpretation of the art installation in Tecate, Mexico called Kikito, part of the series Giants by French artist JR (2017) in which a Mexican child seems to be peering over the wall on the US-Mexico border (http://www.jr-art.net/projects/giants-border-mexico). Several students interpreted the size by writing things like, "the size of the picture compared to the size of the wall and the portrayal of the boy's genuine curiosity are what transcend cultural barriers, as well as physical ones" (Oscar, Essay 4). Another student interpreted the artist's inclusion of a

> *bebé mirando por encima de la pared fue para que la gente vea que las decisiones de inmigración que se toman hoy afectan a los bebés y niños más pequeños que son inocentes y no tienen un idea de lo que está sucediendo* [baby looking over a wall was so that people could see how the immigration decisions made today affect innocent babies and children who have no idea what is happening].
> (Brielle, Essay 4)

Students considered other installations that JR created with themes of migration in France as well. Several commented on how JR (2017) "bridged the gap between…two countries that hate each other…and showed how both citizens are more alike than they are different" (Brian, Essay 4). Such comments showed themes of appreciating respectful dialogue and building bonding and bridging relationships.

Recognize Inequities

As students gathered, weighed, and reasoned with evidence (disposition two: inquire), they also frequently demonstrated: (1) Their ability to recognize inequities and power disparities in daily experiences; (2) the historic experiences they read about; (3) and their own positions of power they held because of previous and current inequities. Through their essays and artifacts, they also showed (4) appreciation for the values of human dignity and diversity and (5) the respectful inclusion of marginalized voices.

Many students exhibited their ability to recognize inequities and power disparities while reflecting on the first-course project, 'The DNA Journey and the Danger of a Single Story.' Students realized:

> despite access to adequate academic resources…, single-story narratives can exist accidentally due to indifference, or laziness, or prejudices, or any number of reasons. Maliciously, however, these narratives can also be created intentionally…single-story narratives not only threaten the intellect of someone or a group of people but, as can be seen at the border, threaten people's very lives.
>
> (Jason, Essay 1)

Students drew conclusions from the content about historical and current inequities, relating the "African Americans who fled the Jim Crow South during the great migration" to the "first settlers of the US" since they "were driven by some of the same motives…escaping a government they found corrupt" (Dina, Essay 2). Another student compared African Americans from the Great Migration to Latin American migrants of today, saying, "African Americans struggles within their castes, and now today we have Latin Americans facing the same issue. Here these people are fighting for citizenship in the same place they were born" (Diana, Essay 2). They were also able to reflect on why people migrate by reading and reflecting on letters from Irish immigrants during the potato famine, writing, "Stressful, even deadly conditions aren't desirable and humans have a natural desire to escape that situation" (Kaden, Essay 3). Some were able to understand migration as excruciatingly difficult through these letters, reflecting, "It was also a very unfortunate time in America for Irish Immigrants since they were seen as inferior humans, and no one would hire them" (Dina, Essay 3).

As they were reflecting on the material through thinking routines, some also realized the role art has in including marginalized voices, recognizing how JR (2017) "wanted to bring light to the fact that instead of dividing us further, we should be accepting and inclusive as humans" (Blake, Essay 4), including the power of incorporating children's marginalized voices since "[p]eople feel more remorse toward children and it sparked empathy in people. I think he made this choice to show how our immigration reforms are affecting the innocent children of other countries" (Kim, Essay 4).

Take Action toward Inclusive and Sustainable Societies

In demonstrating their ability to *take action toward inclusive and sustainable societies*, students reflected on a desire or ability to get involved, including: (1) A sense of belonging to a learning environment and a desire to participate in issues regarding migration; (2) being sensitive toward opportunities; (3) employing understanding, voice, and capacity for influence; and (4) a repertoire of tools for civic engagement. Some students demonstrated this disposition by no longer feeling embarrassed by their culture as when they were a child, but rather "embracing [it] and displaying it for all to see" (Alex, Final Reflection). Others reflected on the importance of discussing immigration rather than avoiding the topic during the conversation, reflecting:

> I learned that nothing gets achieved by not talking about it. In order to solve this crisis we have at the border, people need to talk. Unfortunately, we are so polarized right now that if you say you are on one side or the other, people will ridicule you and shut you down and don't even want to listen. This crisis requires people to be able to understand what the other side is actually worried about and work to find common ground. Immigration is never an easy thing, but it's a necessary thing for our nation.
>
> (Andrew, Final Reflection)

Another student echoed, "we need to stop looking at things from our own perspective but sit down and really listen to what people have to say" (Jaxon, Final Reflection). Finally, another student expressed the importance of learning about immigration in class

> because it is a very hot issue now. It's important that people understand that immigrants aren't coming to the united states to 'Steal your Job' or to sell drugs, a vast majority of them just want to have a decent quality of life.
>
> (Elliot, Final Reflection)

Doing so through historical, multiple perspectives, rather than focusing solely on the current migration crisis, as well as implementing thinking routines allowed students to consider other perspectives and connect historical and present-day stories and inequalities they may never have encountered previously. Furthermore, doing so in the students' FL facilitated their use of Spanish in additional contexts. Repeated use of thinking routines such as those used here (Table 6.1), and identifying them as such, may allow students to use them in the future when considering complex topics.

Challenges & Lessons Learned

Evidence of all five core dispositions suggests that the materials students analyzed and interpreted, the thinking routines they used, and the final course projects they created were a successful combination, as measured by the five dispositions, offering a high impact, learner-centered pedagogy in which SDG 4.7 can be achieved. Furthermore, many students reflected on having new perspectives about the value of migration and a greater appreciation of cultural diversity and of cultures' contributions to development. However, not all student reflections and comments demonstrated equally balanced dispositions;

that is, components of the dispositions were present (sensitivity, inclination, and ability) but may need further exploration to reflect dispositional behavior fully. In fact, as part of the reflection process, some student comments contained strong themes of white privilege or white superiority, minimization of the suffering and hardships that many migrants suffer and continue to experience, as well as assumptions or generalizations of their lived experiences. In all cases, students expressing these ideas also expressed other, more open dispositions in the same reflection. For example, one student wrote in his final paper, "It makes me think that immigration is a very beautiful process to go through. It may have some bumps along the journey to success, but in the end, it is all worthwhile" (Michael, Final Reflection), minimizing the hardships immigrants go through, while just two sentences later writing, "If there is one thing I do want to see is a resolution where the process of immigration is legally followed and followed on no stereotypes towards culture whatsoever" (Michael, Final Reflection). Therefore, it is important to be aware that such reflections and comments are likely to occur as part of the learning process and the lack of awareness of the lived experiences of others. Is it possible to really see migration as a contributor to sustainable societies, in connection to other sustainable development goals such as SDG 10: Reduce inequalities, or SDG 16: Peace, Justice, and Strong Institutions, without also addressing white supremacy? Knowing these comments are likely to occur should not prevent educators from implementing this type of sustainability-based curriculum in the classroom; instead, it is all the more reason to continue encouraging sustainability-based education across the curriculum. Using additional thinking routines to explore complexities and develop perspective-taking could be one way to address such challenges, since thinking routines, as explained by Hammond (2015), allow learners "to monitor and evaluate their comprehension" (p. 132) when coupled with metacognitive or self-regulation strategies, because they "loosely guide learners' thought processes…and facilitate the accomplishment of specific goals or tasks" (Project Zero's Thinking Routines Tool box, n.d.b.).

Concluding Remarks

In a globalized world with unresolved social, political, economic, and environmental challenges, inclusive and equitable quality education that helps build peaceful and sustainable societies is essential (UNESCO, 2016, p. 49). We strongly believe there is an urgent call to include sustainable development topics within broader contexts such as migration, environmental degradation, natural disasters, etc. in the FL classroom, since learning a language should involve the development of knowledge, skills, values, and attitudes required to develop intercultural understanding to create more equitable societies.

The project we have presented here is no exception. Materials for the project are mainly open-access and are not tied to a single textbook or curriculum; the SDGs can inform curriculum design and programming in many different ways; thinking routines can be implemented across the curriculum in any language, with or without the incorporation of technology, and can be essential in discussing a polarizing topic in a state which is detaining countless migrants and separating children from parents in detention centers along its southern border.

A global perspective through SDGs, in addition to strong thinking patterns and discourse organized through thinking routines, as well as a historical exploration of content, all combined proved to be a catalyst for a humanistic learning process in the Spanish language classroom. Therefore, it behooves us to incorporate additional CBI into our curriculum and pedagogies. Although students may resist a radical pedagogical approach to language learning, especially as part of a core requirement course, our ultimate goal is to promote a more sustainability-based, content-enriched pedagogy so they will hopefully no longer consider the Spanish Foundations coursework just a requirement they need to fulfill, but rather, an opportunity to build language and sustainability literacy (for SDG 4.7 and beyond), as well as long-term learning and thinking strategies for success and global participation.

Notes

1 The 2015 World Education Forum was held in Incheon, the Republic of Korea. The goal was to set a new educational vision for 15 years after the Forum (UNESCO, 2016).
2 The courses are taught almost exclusively by graduate students in Spanish.
3 In this chapter, migration is explored as a context to achieve SDG 4.7.
4 The Re-Imagining Migration framework is a multi-faceted educational repository that offers resources and materials including, but not limited to, Thinking Routines and Dispositions for a World on the Move, a learning arc for curriculum design and implementation related to migration, and civics, inquiries, lessons, and resources.
5 A disposition is a psychological element with

> three components: inclination, sensitivity, and ability. [...] Inclination refers to the person's felt tendency toward behavior X. Sensitivity, in contrast, refers to the person's alertness to X occasions [...] and ability refers to the actual ability to follow through with X behavior.
>
> (Perkins et al., 1993)

6 To conduct the interviews, students are directed to the Moving Stories Project on the Re-imagining Migration website, in which a collection of questions is provided to explore migration stories, as well as tips for conducting interviews successfully: https://reimaginingmigration.org/moving-stories-home/
7 In July 2020 Veronica Boix-Mansilla revised the Dispositions for a World on the Move Framework. In this version, the five dispositions organizing our analysis and exposition remain unmodified, lending sustained grounding to our analysis. Only sub-items were modified within each disposition to diminish repetition, enhance clarity in language, and sharpen the articulation of cognitive and socioemotional dimensions of learning. This revision was published online in January 2021 (V. Boix-Mansilla, personal communication, May 12, 2021).
8 All names are pseudonyms.
9 Excerpts are inserted as students wrote them and have not been edited. The authors have translated those written in Spanish.

References

ACTFL. (2012). *ACTFL proficiency guidelines*. American Council on the Teaching of Foreign Languages. https://www.actfl.org/sites/default/files/guidelines/ACTFLProficiencyGuidelines2012.pdf

Agency by Design. (2018). *Voice and choice*. http://www.pz.harvard.edu/sites/default/files/Voice%20and%20Choice_1.pdf

Babaci-Wilhite, Z., Geo-JaJa, M., & Lou, S. (2012). Education and language: A human right for sustainable development in Africa. *International Review of Education, 58*(5), 619–647. https://doi.org/10.1007/s11159-012-9311-7.

Badjanova, J., & Iliško, Dz. (2015). Holistic approach as viewed by the basic school teachers in Latvia. *Discourse and Communication for Sustainable Education, 6*, 132–140.

Boix Mansilla, V. (n.d.). *Five dispositions for a world on the move*. https://reimaginingmigration.org/dispositions-for-world-on-the-move/.

Cammarata, L. (2016). Foreign language education and the development of inquiry-driven language programs: Key challenges and curricular planning strategies. In L. Cammarata (Ed.), *Content-based foreign language teaching: Curriculum and pedagogy for developing advanced thinking and literacy skills* (pp. 123–146). New York: Routledge.

Cammarata, L., Tedick, D. J. & Osborn, T. A. (2016). Content-based instruction and curricular reforms: Issues and goals. In L. Cammarata (Ed.), *Content-based foreign language teaching: Curriculum and pedagogy for developing advanced thinking and literacy skills* (pp. 1–21). New York: Routledge.

de la Fuente, M. (2019). Stepping out of the language box: College Spanish and sustainability. In C. A. Melin (Ed.), *Foreign language teaching and the environment: Theory, curricula, institutional structures* (pp. 130–145). New York: Modern Language Association.

Hammond, Z. (2015). *Culturally responsive teaching and the brain: Promoting authentic engagement and rigor among culturally and linguistically diverse students*. California: Corwin Press.

Jodoin, J. (2019). Promoting language education for sustainable development: A program effects case study in Japanese higher education. *International Journal of Sustainability in Higher Education, 21*(4), 779–798. https://doi.org/10.1108/IJSHE-09-2019-0258.

JR. (2017). *Giants, Kikito*. Retrieved March 18, 2021, from http://www.jr-art.net/projects/giants-border-mexico

Lantolf, J. (2000). *Sociocultural theory and second language learning*. Oxford: Oxford University Press.

Leaver, B., & Stryker, S. (1989). Content-based instruction for foreign language classrooms. *Foreign Language Annals, 22*(3), 269–275.

Martel, J. (2016). Tapping the national standards for thought-provoking CBI in K–16 foreign language programs. In L. Cammarata (Ed.) *Content-based foreign language teaching: Curriculum and pedagogy for developing advanced thinking and literacy skills* (pp. 101–121). New York: Routledge.

Melin, C. A. (Ed.). (2019). *Foreign language teaching and the environment: Theory, curricula, institutional structures*. New York: The Modern Language Association.

Met, M. (1998). Curriculum decision-making in content-based second language teaching. In J. Cenoz & F. Genesee (Eds.), *Beyond bilingualism: Multilingualism and Multilingual Education* (pp. 35–63). Philadelphia: Multilingual Matters.

https://doi.org/10.1632/prof.2007.2007.1.234.

Perkins, D. (2003). Making thinking visible, accessed June 6, 2020, available at http://www.pz.harvard.edu/resources/making-thinking-visible-article-0.

Perkins, D., Jay, E., & Tishman, S. (1993). Beyond abilities: A dispositional theory of thinking. *Merrill-Palmer Quarterly, 39*(1), 1–21. https://www.jstor.org/stable/23087298.

Pevneva, I., Gavrishina, O., Smirnova, A., Rozhneva, E., & Yakimova, N. (2017). Education within sustainable development: Critical thinking formation on ESL class. *E3S Web Of Conferences, 21*, 1–5. https://doi.org/10.1051/e3sconf/20172104021.

Project Zero. (n.d.a). *Re-imagining migration.* https://pz.harvard.edu/projects/re-imagining-migration.

Project Zero. (n.d.b). *Project Zero's thinking routines toolbox.* https://pz.harvard.edu/thinking-routines.

Project Zero's Thinking Routines Toolbox. (2019a). *Connect, extend, challenge.* http://www.pz.harvard.edu/sites/default/files/Connect%20Extend%20Challenge_0.pdf

Project Zero's Thinking Routines Toolbox. (2019b). *Circle of viewpoints.* http://www.pz.harvard.edu/sites/default/files/Circle%20of%20Viewpoints_0.pdf

Project Zero's Thinking Routines Toolbox. (2019c). *The 3 whys.* http://www.pz.harvard.edu/sites/default/files/The%203%20Whys.pdf

Re-imagining Migration. (n.d.). *Framework.* https://reimaginingmigration.org/a-framework-of-opportunity/. Retrieved November 7, 2020 from https://reimaginingmigration.org/a-framework-of-opportunity/

Ritchhart, R., & Church, M. (2020). *The power of making thinking visible: Practices to engage and empower all learners.* Hoboken: Jossey-Bass.

Stoller, F. (2004). Content-based instruction: Perspectives on curriculum planning. *Annual Review of Applied Linguistics, 24*, 261–283. https://doi.org/10.1017/S0267190504000108.

ter Horst, E., & Pearce, J. (2010). Foreign languages and sustainability: Addressing the connections, communities, and comparisons standards in higher education. *Foreign Language Annals, 43*(3), 365–383.

Thomas, P., & Thomas, D. (2002). *What is an Accordion Book.* Baymoon.com. Retrieved 3 December 2020, from http://www.baymoon.com/~peteranddonna/5-Definition%20of%20Accordion%20Book.htm.

UNESCO. (2016). *Incheon declaration and framework for action for the implementation of sustainable development goal 4: Ensure inclusive and equitable quality education and promote lifelong learning opportunities for all.* UNESCO. Retrieved 3 December 2020 from https://unesdoc.unesco.org/ark:/48223/pf0000245656.locale=en.

Willingham, D. T. (2019). *How to teach critical thinking* (pp. 1–17). Education: Future Frontiers.

Zygmunt, T. (2016). Language Education for Sustainable Development. *Discourse and Communication for Sustainable Education, 7*(1), 112–124. https://doi.org/10.1515/dcse-2016-0008.

7 Toward the Greening of the Intermediate French Language Curriculum

Richard Kern and Vesna Rodic

Introduction

The Anthropocene Epoch may be an unofficial and controversial designation of our current period of planetary evolution, but it reflects the indisputable truth that human activity has had a significant impact on our planet's climate and ecosystems. We are living in what the 2020 United Nations Human Development Report calls "a mass species extinction event...caused by a single organism—us" (Conceição, 2020, p. 3). The environment and its preservation are understandably a preoccupation of young people today, for they will be the ones to live with the consequences of inadequate response at this critical juncture. Environmental sustainability, because of its global urgency and inherent interest to students—and because it is a concept linked to key developments throughout French cultural history—became a natural theme for a pilot project in our second-year French curriculum at the University of California at Berkeley. This chapter describes this pilot project, whose goal was to introduce students of Intermediate French to a content-based study of primary sources related to environmental sustainability at UC Berkeley and in French-speaking contexts. The curriculum incorporated activity sequences involving legal documents, literary works, visual images, maps, and films to study a range of environmental issues, both contemporary and historical. The pilot project also involved videoconferencing exchanges with Master's degree students from the Université Aix-Marseille (AMU) in France who were preparing for careers in French language teaching. This chapter also describes some of the curricular changes implemented following the pilot project.

Before going into the details of the pilot project, we must first establish the importance of the environment in French culture and discuss the pedagogical rationale that guided our decisions. Study of the environment has been a component of primary and secondary education in France since 1977. France was the first country to add a Charter on the Environment to its Constitution in 2004, and in that same year, the national curriculum was expanded to include sustainability. Universities receive ratings for their sustainable practices, and secondary schools have appointed eco-delegates. France emerged as a world leader in sustainability efforts

DOI: 10.4324/9781003080183-10

during the 2015 Paris Accord on climate change (United Nations Climate Change, 2021). Of particular importance for language learners is that environmental themes have long been represented across various genres in French literature, from the 16th century to the contemporary period. Environmental themes are also present in French films and visual arts and are deeply woven into the French psyche. For the Berkeley students, then, the study of French texts, art, and regional case studies of environmental action was designed to motivate them to develop an 'insider's perspective' into French culture. For their exchange partners at AMU, the goal was partly to discover Californians' practices and perspectives, and principally to gain experience in designing learning opportunities for students of French language and culture.

Given the significance of environmental themes in French literature, history, arts, politics, and contemporary life, we determined from the outset that a content-based approach (Cammarata, 2016; Krueger & Ryan, 1993; Stoller, 2002) would provide optimal support for students' learning because it cultivates learners' ability to work with primary sources, to engage in experiential learning, and to think comparatively and transculturally (MLA Ad Hoc Committee on Foreign Languages, 2007). This unified language-and-content curricular approach was recommended by the MLA (Modern Language Association) Ad Hoc Committee on Foreign Languages (2007) to reconfigure college foreign language programs. Simpson and Mueller (2019) add that a content focus on the environment means moving "away from an anthropocentric worldview toward biocentrism" and that "[t]his reconfiguring of the foreign language curriculum insists on environmental literacy as a significant learning outcome alongside proficiency in the target language; it adds environmental literacy to more conventionally conceived cultural competence" (p. 19).

Some projects that have addressed sustainability in French curricula include Sampon-Nicolas (2019) and Carrico (2019). Sampon-Nicolas introduced her students to what she calls "sustainability literacy" in French literature and film, engaging her students with questions of *ecological identity* (p. 46) and *environmental awareness* (p. 48), culminating in their writing of a proposal for a microfinance project and a multi-genre research project. Through these experiences, her students "come face-to-face with how present-day problems relate to personal choice... [and become] aware of how other cultures confront similar issues" (p. 51). Carrico (2019) developed a 19th-century French literature course that gets students to consider how 19th-century French representations of water intersect with modern-day problems and how past perceptions are maintained or challenged today. Carrico points out that the crucial element of the course is its structure, designed "to help students situate changing thought and make comparisons between genres and representations" (p. 91). These structural goals are shared in our project, although ours is distinctive in that it is designed for the lower-division instructional level.

Pedagogical Underpinnings

Our approach in this curricular project is grounded in content-based instruction (CBI). The idea behind CBI is that people can learn language incidentally through the process of studying some content matter other than language itself when that content is presented in the target language. It is a student-centered approach based on the notion that learners' motivation is enhanced when they work with stimulating content. This motivation facilitates language learning, which, in turn, makes more content material (and learning) accessible to them. CBI relies primarily on authentic language materials. Stoller (2002) underlines the reciprocal relation that characterizes content-based approaches, whereby "language [serves] as a medium for learning content and content [serves] as a resource for learning and improving language" (p. 109). In practice, however, 'content-based' can be quite variable in the balance of emphasis on language versus non-language subject matter learning (Lightbown, 2014; Met, 1998; Snow, 2014).

Within this overarching approach, our pilot and the subsequent changes to our intermediate French curriculum were guided by three major principles. The first was the use of what we call a *discovery/inquiry-based* pedagogy. This can be unpacked as follows. Since 2017, the UC Berkeley campus has made 'discovery' a new framework for the undergraduate student experience. Discovery has been articulated as a four-part process—connect, discover, engage, reflect—in intentional and immersive educational experiences that allow students to explore, develop, and extend knowledge beyond the self (Discovery Work Group, 2021). We describe below how we have integrated these processes pedagogically. Inquiry learning (Hmelo-Silver et al., 2007) emphasizes asking questions, collecting and analyzing data, and making evidence-based arguments. In our project, we use the term to refer to the Library of Congress's initiative to help teachers encourage their students to engage with primary sources. This involves comparing multiple sources that represent differing points of view, dealing with contradictions, and confronting complexity as students ask questions about the origins of source material, the contexts(s) of use, authorial intentions and rhetorical choices, and biases and stereotypes (Library of Congress, 2021). This approach emphasizes a cyclical three-stage process of questioning, observation, analysis, and more questioning—in our case, applied to learning about policies, sustainable practices, and attitudes about environmental sustainability in both France and the US. Another way we oriented students toward inquiry was by having them engage with graduate students in France via regular videoconference exchanges (to be described below). The goal of these exchanges was to promote critical dialogue with French peers, with hopes of generating new transcultural insights and a potential for social action.

The second principle, closely related to the first, was integrating *primary sources* and *experiential learning* to ground students' knowledge in direct experiences that would contextualize information. Primary sources challenge students to think, imagine, and explore. The inquiry-based methods described above were applied to the analysis of primary sources by introducing students to documentary research in libraries and online, examining a variety of primary texts using the US Library of Congress's tools for the analysis of primary sources (Library of Congress, 2021), as well as local community research (e.g. at the Berkeley Food Institute and the Basic Needs Center, as well as in students' local communities) involving observation, interviews, and other forms of data collection. Such experiences were inspired by principles of experiential learning, which integrates knowledge, activity, and reflection. Experiential learning derives from Dewey's (1897, 1963) insight that all learning occurs within a social environment and that the educator's task is to facilitate appropriate experiences that will actively engage students. Kolb (1984) proposed a four-part cycle of experiential learning that involves moving from (1) concrete experience to (2) reflective observation to (3) abstract conceptualization to (4) active experimentation. This process involves initiative, creativity, decision-making, problem-solving, and meaning construction, and is highly compatible with the Library of Congress's tools for the analysis of primary sources mentioned above.

The third principle was the creation of discovery pathways and research connections, following the spirit of the interschool exchanges endorsed by Célestin Freinet (1948, 1994) and more recently represented by telecollaboration projects such as *Le Français en (première) ligne* (Develotte et al., 2007).[1] By incorporating a distance dimension via telecollaboration, our project aimed, as Cummins and Sayers (1995, p. 137) describe, to enhance learners' awareness of the social, cultural, historical, geographic, and linguistic realities of their own community by their need to describe these realities in response to their distant partners' questions. *Distance* thus involves more than geographical separation but also a critical disposition to reflect on differences in meaning, mentalities, and worldviews, along the lines of the goal promulgated by the 2007 MLA Report that learners should "reflect on the world and themselves through the lens of another language and culture... to grasp themselves... as members of a society that is foreign to others" (2007, p. 4). At the same time, we aimed to highlight diversity *within* cohorts, recognizing that in a globalized society, we are all increasingly pluricultural individuals.

The Project: Environmental Sustainability in the Intermediate French Curriculum

During the 2018–2019 academic year, 128 intermediate (second-year) French language students at UC Berkeley participated in a pilot project focused on the theme of the environment. In the Fall, 29 students across two sections of third-semester French participated, as did 15 students from one section of fourth-semester French. Another 84 students participated in the second semester (54 from French 3, and 30 from French 4). Supported by a Library of Congress Teaching with Primary Sources grant,[2] this project aimed to introduce students of Intermediate French to a discovery and inquiry-based approach that engaged them in documentary research and community involvement through visits and interviews with local environmental groups and services, which they presented to their exchange partners in France. For the Berkeley students in our pilot, the global learning objectives were to develop an ability to engage in authentic cultural discussions, write analytically about texts, compare cultures—considering multiple perspectives and thinking interculturally—, and learn how to use technology effectively and critically. The content learning objectives were to learn about environmental policies, sustainable practices, environmental representations in literature and film, and environmental attitudes in both the US and France. Needless to say, French language development was interwoven with all these goals. We describe below how we used the thematic content of environmental sustainability to support students' French language proficiency and their understanding of environmental sustainability in both local and French contexts and how environmental issues might be framed and thought about differently in different cultures.

Besides the global urgency of the environmental theme and its pertinence to French culture past and present, our project also addressed a need to provide continuity concerning the theme of the environment within the French language curriculum at UC Berkeley. Our second-semester (elementary) French course devotes one unit to introduce students to sustainability content in French by including a text about the environmental crisis in a lesson titled *La planète va mal* ('The planet is not feeling well'). Although this unit presents essential French vocabulary and expressions related to the environment (such as recycling, carbon footprint, sustainability, etc.), it frames the theme of the environment mainly through a political lens, as the chapter ultimately asks students to get involved in a political sense (*engagez-vous*) and to create a poster on one environmental issue, showcasing a particular problem and presenting efforts towards its solution. While this unit provides a valuable first encounter with the ecological theme, it oversimplifies a complex and essential aspect of French culture. For instance, students are presented limited information about selected eco-friendly practices in France but are not asked to discover much about them independently.

At the time, in the second-year (intermediate) courses, French 3 and 4, the theme of the environment was notably absent from the textbook and was only referenced a few times in passing and without any conceptual, informational, or cultural development. In the interest of improving articulation between our first- and second-year programs as well as the subsequent gateway course to the upper-division, French 102, we reflected on the possibility of getting students truly involved, *engagés*, through a deeper exploration of the theme of the environment both in and outside the classroom. We decided that an inquiry-based approach would best suit this kind of active learning because it places questioning at the heart of the learning process. It also held the promise to bring students closer to an insider's perspective on the target culture. Through concrete, lived experiences of situations and practices related to the subject of environmental sustainability enabled by experiential learning, students would be able to engage in a cross-cultural examination of concepts and activities and consequently to access cultural values of the target culture more profoundly.

The specific focus of this pilot project was to heighten students' awareness of environmental sustainability issues on an international scale, to train them in the use of primary sources, and to provide opportunities for them to explore sustainability and primary sources in both local and French-speaking contexts, engaging with faculty and students at AMU via a series of videoconferences, as well as with UC Berkeley's Office of Sustainability and Energy. The pilot program incorporated activity sequences involving legal documents, literary works, visual images, film, and maps to study a range of environmental issues, both contemporary and historical. This focus on environmental issues would contribute to a deep and meaningful engagement with the target culture, promoting students' intercultural and even transcultural awareness (MLA Ad Hoc Committee on Foreign Languages, 2007).

Curriculum Overview

Three instructional units were developed: two were integrated into the third-semester French course, and one in the fourth-semester French course. The first one, involving the study of a legal document, France's Charter on the Environment, was introduced in week eight of the semester and involved a six-day sequence. The second unit, on food sustainability, was initially introduced in week 13 of the semester.[3] This unit spans three days of in-class work. It is paired with experiential learning on campus and videoconferencing exchanges with students from AMU during the second half of the semester. The third unit involves a comparative study of environmental sustainability as represented in films. It is introduced in week nine of the semester, and it lasts five days. The remainder of the chapter will focus on the first unit (on France's Charter on the Environment) to briefly illustrate how our goals were translated into instructional practice.

Sustainability in Intermediate French: *La Charte de L'Environnement*

France's Charter on the Environment was created in 2004 and added to the French Constitution in 2005, formalizing the country's official stance vis-à-vis the environment. In creating this unit, we supplemented a chapter on politics already in the course in which environmental questions were considered only in passing as one of several topics. This allowed us to introduce environmental topics from a political perspective and then enhance and expand them into a study of environmental efforts through historical and cultural lenses. That is, starting from the study of a political issue—engagement in questions related to environmental sustainability—we designed a lesson sequence that also relied on a discovery of moments of French cultural history (see Table 7.1).

Table 7.1 Overview. Module on *La Charte de l'environnement*

Day	Objectives/approach	Materials/activities	Production/assessment
1	What are the main environmental threats today? (re)access vocabulary of the environment [language development; schema activation] Chronology behind the birth of the Charter of the Environment. [content knowledge development]	Discuss Handout #1 (Chronologie et contexte)	In-class: Freewriting exercise— *Selon vous, quelle est la question environnementale la plus grave aujourd'hui?* Homework: Research the 3 documents that the Charter on the Environment is based on. What kinds of documents do they represent? Then, research the historical context of this document (a) in France and (b) in the US.
2	Discuss homework findings. Textual analysis (3 excerpts from documents referenced by the *Charte*: *La déclaration des droits de l'homme et du citoyen;* The preamble to 1946 French Constitution; The opening paragraph of 1958 Constitution). [primary source analysis, comparison]	In class. Excerpts from the 3 documents (Handout #3). Analyze language, similarities, differences in light of the information about the context that the students had researched. Point out differences between primary and secondary sources.	Homework: Read *La Charte de l'environnement*. Fill out the Analysis Tool Handout #4 (observe-reflect-question steps).
3	Discussion: The Charter on the Environment based on students' questions from Analysis Tool. [inquiry; content knowledge development]	Text: *La Charte de l'environnement* (Handout #2). Class discussion as a whole; group work to address student questions.	Homework: Journal assignment—Read Article 5 of the Charter closely and reflect on its message: In what sense is it controversial?
4	Continue discussion on the Charter on the Environment. Engage in close textual analysis. [inquiry; primary source analysis; content knowledge development]	In-class discussion (include homework responses from the journal). Consider values that the Charter conveys: Do a word map.	Homework: Examine members of the Coppens Commission (Handout #1): what professions are represented, what is their value? How does the make-up of this Commission influence your view of the Charter?

| 5 | Discussion: Context surrounding creation of the Charter. Cross-cultural examination of distinct standpoints towards ideas expressed in the Charter. [inquiry; content knowledge development; primary source analysis; language development] | In-class discussion and role-play, examining the standpoint of each member of the Committee (how profession might influence attitudes toward the Charter). | Homework: Prepare for Debate—The French Parliament is voting on the Charter. What do members embrace, question, contest? Why? Pick a *Département*; you are its representative. Create a photo- essay on the region's characteristics, cast a vote and explain reasons for your *Département's* vote. |
| 6 | Engage in cross-cultural, inquiry-based reflection on distinct attitudes toward the environment. [inquiry; cross-cultural reflection; language development; experiential learning] | In-class debate. | Photo-essays. |

Day 1

Students in the class were asked to reflect, in writing, about the most press-ing environmental question today and their responses showed two common issues: water and air pollution. This exercise allowed students to revisit their knowledge of the environmental theme, re-activate environmentally-specific vocabulary, and engage in expository writing in French, all of which served as preparation for analyzing the primary document, *La Charte de l'environnement*. Before they looked at the *Charte*, they needed to become familiar with the chronology behind the birth of the Charter in France by researching infor-mation on the three chief documents that the Charter on the Environment references: The Declaration of the Rights of Man and Citizen from 1789, the Preamble to the French Constitution from 1946, and the opening lines of the French Constitution of the Fifth Republic from 1958, still in effect today. Stu-dents were asked to consider the type of documents these works represented and research cultural information on the years they were produced. As they engaged in this exploration of French cultural history, they were encouraged to think comparatively and reflect on events taking place in the US during the same years. Students discovered information on the French Revolution, the Second World War, and French colonization through the research process.

Day 2

In the following class session, in-class analysis of excerpts from the three pri-mary documents that influenced the *Charte* allowed students to practice active reading and comparative analysis of the three sources. Students identified key

FICHE D'ANALYSE POUR SOURCES PRIMAIRES

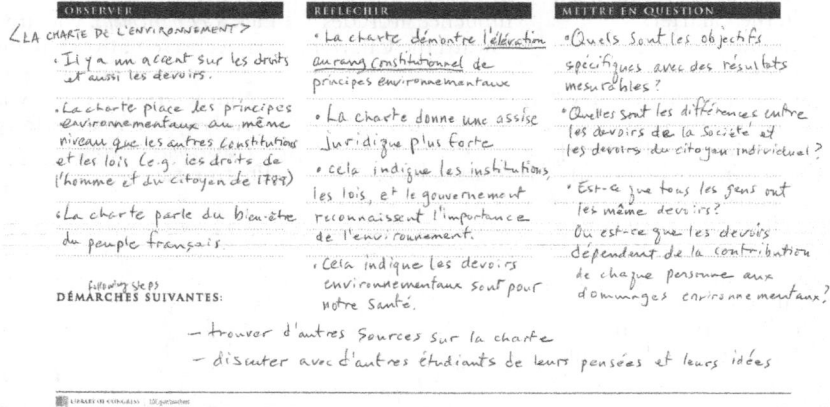

Figure 7.1 Sample student response to the Library of Congress Primary Source Analysis Tool. Courtesy of Library of Congress. Using primary sources. Primary Source Analysis Tool.

terms, noted similarities and differences between the three texts, and considered the documents' intended audience and message. Finally, for homework, students read the Charter on the Environment itself and filled out a Primary Source Analysis tool, provided through the Library of Congress Teaching With Primary Sources website, asking students to organize their reading experience in three major steps: 'Observe,' 'Reflect,' and 'Question' (see Figure 7.1). The worksheet additionally prompted students to consider 'Further Investigation' to individualize their research path. Having completed the initial reflection about the text, students were encouraged to think about other information they wanted to know about and consider possible avenues of research that could help them acquire answers.

Day 3

Students were able to identify several important segments of the text, as well as its key terms and phrases, in the 'Observe' column, noting especially the relevance of the terms 'rights' [*droits*] and 'responsibilities' [*devoirs*]. In the 'Reflect' column, students further examined the textual excerpts that they had identified. For example, a student who noted "everyone has the duty to contribute" as a key term in the 'Observe' column elaborated in the 'Reflect' column that "the proclamations are a bit ambiguous and general." The third column ('Question') gathered various questions that the students generated about the text. These ranged from vocabulary and general comprehension queries to analytical questions related to the content of *La Charte de l'environnement* to broader questions about humankind's relationship to nature. Some students asked about the French cultural context in order to compare it with the US situation. For example, students wondered about how many French people engaged in recycling, whether the French school system accounted for the preservation of the environment, and what actions the French government took concerning sustainability. Students wondered about the lack of a Charter in the US, and contemplated distinctions between the French and American legal systems. Students further inquired about the relationship between *La Charte de l'environnement* and the Paris Climate Accord, as well as the importance of the texts that inspired *La Charte*. With the scaffolded access to the *Charte* through the Primary Source Analysis Tool, students were able (a) to engage in both a close reading of the text itself and a broader reflection on the concepts that it indexed, and (b) to situate the *Charte* in relation to various moments in French cultural history, both past and contemporary. In the journal assignment that followed as homework, students analyzed Article 5 of *La Charte*, which requires public authorities to assess risks and adopt protective measures when human actions are suspected of causing environmental damage. In their journals, students made significant connections between various parts of *La Charte*.

Day 4

Looking at a single article enabled both an exercise in textual analysis and a reflection about the implications behind the document. For instance, students were able to relate the document's key concepts of roles and responsibilities toward nature to a reflection on the role of the French State in individual actions. The module culminated with an in-class debate in which students engaged in role-playing and were asked to assume a role of an assemblyman, preparing to vote *for* or *against* the Charter on the Environment in 2005. Students were randomly assigned a *Département* (an administrative sub-division of a region in French territories). They were asked to research historical, economic, and political information regarding that *Département* and then decide whether their constituents would favor *La Charte* based on the region's characteristics, financial products, and constituents' needs and interests.

Day 5

This day focused on continued discussion of *La Charte* and preparations for the debate. The debate involved creating a photo-essay containing a brief description of the region and an analysis of how the geographical parameters and the economy of a *Département* might affect its decision to vote for or against *La Charte de l'environnement* (see Figure 7.2). Thus, as seen in the example of the Alpes-Maritimes *Département*, students considered the relationship between

Figure 7.2 Student slides describing the Département des Alpes-Maritimes

the *Département*'s natural elements and tourism, its benefits for the region's economy, and the environmental consequences of climate change on local ecosystems such as ice melting, disappearing lakes, heat waves, floods, and forest fires. In this way, students engaged in a detailed discovery of the features of the *Département* and were able to delve deeply into critical environmental issues that would inform their vote on *La Charte*.

Students showed great interest in the role-play dimension of the debate. They found that the Analysis Tool worksheet helped structure their reading and reflection processes and even improve their basic understanding of the text. As students organized their thoughts about the *Charte de l'environnement*, the kinds of questions they generated through the inquiry-based approach enabled the class to engage more deeply with the primary source's cultural context.

Day 6

The in-class debate occurred on Day 6. Students engaged in role-playing as representatives of specific *Départements* in the French Assembly. Students had prepared slides for which they researched the economic, social, and environmental characteristics of their *Département*, attempting to gauge how its constituents would vote on *La Charte*. Finally, students were asked to cast a vote and formally present arguments for or against, as a deputy might. To prepare, students looked at France's *Assemblée Nationale* (2019) website and watched live proceedings on YouTube. Following each vote cast, students were asked to justify their decision based on the characteristics of the region that they represented.

Through this unit, our intermediate French students gained an in-depth understanding of environmental concepts and their treatment in both

French- and English-speaking contexts, as seen in their written responses on the Primary Source Analysis Tool, their in-class discussion during the debate, and their photo-essays. They engaged in a significant discovery of environmental topics, noting that France's *Charte de l'environnement* relies on rhetorical reasoning to appeal to particular emotions. They expressed interest, for example, in the fact that both farmers and philosophers were represented on the committee that drafted the *Charte*. In the process of analyzing the *Charte*, they considered the social, political, and historical influences that helped shape France's contemporary attitudes toward the environment. Finally, in terms of language development, students expanded their vocabulary related to the environment and foods, deepened their familiarity with the debate genre, and exercised their abilities in argumentative speech.

Analysis of Student Work

In the unit on France's *Charte de l'environnement*, students were particularly prolific in generating various questions regarding the document. As they gained experience with the crafting of analytical questions, students then followed up by creating their own discovery pathways to explore the questions they identified about the Charter. Students' questions ranged from those about vocabulary and close reading to those engaging the historical and cultural context to those comparing French and American societies. Close reading and the work on vocabulary yielded analytical questions involving key concepts expressed in the document. For example, based on her close reading, the student who completed the Primary Source Analysis Tool shown in Figure 7.1 asked "What are the differences between the duties of the society and the duties of the individual?" and "Do all people have the same responsibilities, or do responsibilities depend on the degree to which each individual contributes to environmental degradation?"

Students were interested in specific practices in France that have emerged in response to the Charter on the Environment. For example, three students focused on metrics, asking about the number of people in France participating in recycling efforts (*"Combien de français font le recyclage?"* [How many French people recycle?]), and whether this document had instilled concrete changes in the French society since its creation (*"Comment est-ce que cette charte change l'environnement et la vie pour le peuple français?"* [How does this Charter change the environment and daily life for the French?]; *"Est-ce qu'il a des actions concrètes qui viennent de la charte?"* [Have there been concrete actions that have come from the Charter?]). Another student became fascinated with environmentally conscious stores and the phenomenon of greenwashing. These questions set students on a pathway of discovery around specific environmental issues and enabled them to gather metrics related to recycling.

Several students were interested in governmental and educational practices related to the implementation of the Charter in France, which allowed them to research environmental topics within specific cultural contexts. For instance, one student asked about the link between environmental efforts and France's education system (*"Est-ce que le système scolaire donne la priorité à l'éduca-tion de la préservation de l'environnement?"* [Does the school system give priority to environmental education?]), while another student focused on specific actions that the government takes in the domain of sustainability (*"Quelles actions est-ce que le gouvernement prend sur la durabilité?"* [What actions does the government take concerning sustainability?]). These questions further anchored the study of the target culture within specific cultural contexts. They opened up the theme of the environment around cultural values, seen in the attitudes and practices that our language students further encountered on their individual discovery paths.

Students were further able to connect the document to other moments in recent French cultural history and thus explore the contemporary context. For instance, two students asked about the reasons behind the creation of the Charter (*"Pourquoi est-ce que ce document est écrit à 2004?" "Quel événement s'est passé à cette fois?"* [Why was this document written in 2004? What event was hap-pening at this time?]). One student wondered about the possible relationship between the Charter and the Paris Climate Accord from 2015 (*"Comment la Charte de l'environnement de 2004 a influencé la création de l'Accord de Paris?"* [How did the Charter of the Environment of 2004 influence the creation of the Paris Accord?]), while another student was interested in the Charter's influence in the present day (*"Est-ce que ce document a le même importance aujourd'hui?"* [Does the document have the same importance today?]).

Most students compared France's sustainability efforts with those in the US. One student wondered: *"Pourquoi est-ce que les États-Unis n'a pas une charte comme ça- Qu'est-ce que est différent en France qu'on pouvait créer cette charte?"* [Why doesn't the United States have a charter like this one- What is different in France that it was possible to create this charter?]. Another student asked: *"Pourquoi la France a reconnu la réalité de les problèmes environmentale comme une prob-lème légale, et pas les États Unis?"* [Why has France recognized the reality of environmental problems as a legal problem, and not the United States?]. An-other student focused the question on differences in legal systems between France and the US, asking: *"Est-ce que cette système légale influence les Françaises de penser plus d'environnement plus de les Américains?"* [Does this legal system influ-ence the French to think more about the environment than the Americans?]. These questions allowed students to engage in cross-cultural and even trans-cultural thinking (MLA Ad Hoc Committee on Foreign Languages, 2007), to examine a variety of viewpoints on the environment, and to gauge possible differences between them.

In responses to assessment questionnaires administered after the pilot and subsequent curricular revisions, students of French were particularly passion-ate about the *experiential* components of learning: Telecollaboration with stu-dents from AMU and projects on campus and in their local communities. Overall, students underscored the value of the *practice* of research skills and academic inquiry (e.g. "practice in handling primary sources"; "practicing this in class [helped]") as well as the added benefits of conducting research *in another language*: "Doing the research for my cultural presentation entirely in French allowed me to practice a kind of research that I don't usually do."

Content-Based Instruction and a 'Greener' Intermediate French Curriculum

As shown in the content-based unit *La Charte de l'environnement*, our instructional approach combined the study of primary sources with inquiry-based learning and experiential learning to broaden the scope of the study of environmental themes. We have adopted this approach in other parts of our second-year curriculum, building on the work done in our pilot project. In third-semester French, in a unit on cultural icons, we have introduced a study of visual and textual materials on the 'imperfect produce' (*les fruits et légumes moches*) movement in France marking the country's leading role in food sustainability today. In our decision to include this material in a unit on cultural icons, we considered the possibility that, rather than products themselves, the *practices and attitudes* towards food and food sustainability are powerful sources of information about French culture and its values. The exploration of this topic was further expanded through videoconferencing exchanges with students from AMU, who provided the Berkeley students with information and researched examples of contemporary food sustainability efforts in France while maintaining an ongoing exchange in French. Our integrated approach to the study of environmental issues has provided a model for discussions of environmental dimensions in units on gender, family, and urban and suburban life, where students have been able to engage with a variety of primary sources, research information at the library or online, and visit sites on campus as well as in their local communities.

In our fourth-semester language course (French 4), students can discuss the theme of the environment in units on film, *Francophonie*, and the arts. The film unit has exposed students to feature films and documentaries and has included visual and film analysis and creative writing. Such an approach has allowed students to assume distinct standpoints concerning the materials by selecting scenes from films to nominate for class discussion, analyzing features of a specific film genre, and adding codas to the films through imaginative role-play. The creative work and role-plays have enabled students to draw from their previously acquired knowledge on environmental questions (and the cultural values reflected in them) to justify their choice of scenes added to the films. Finally, such work has prepared them for subsequent course levels, where the emphasis is placed on analytical writing, the study of primary and secondary sources, and research skills.

Challenges and Recommendations

Some of the challenges we faced in carrying out this pilot project included the coordination of the telecollaboration project with the AMU teachers and students in terms of curriculum (in that the two cohorts had entirely different kinds of courses) and in terms of logistics (a nine-hour time difference meant that only early morning sections of French could be involved on the Berkeley side and that the class in France had to be held late afternoon/early evening, right up to when the university buildings closed).

Another challenge involved convincing colleagues *not* involved in the course that it was all right for students to use English to investigate local sustainable practices within the context of a French language course.

In terms of student work, an initial challenge pertained to providing more systematic feedback on grammatical structures within the context of creative work, research, and group work, since we did not necessarily want to put creativity and language use at odds through frequent error correction. We have addressed this by devising exercises targeting grammatical structures directly connected with the content of the materials from *La Charte* unit. Having the materials on the environment serves as the cultural framework for the study (and testing) of grammatical structures. This solution has proven quite effective, as it has allowed students to integrate and deepen their knowledge of environmental topics and offer personalized answers while using targeted grammatical structures. More generally, we faced the challenge of evaluating students' learning in ways other than the work they produced in the class and their reflections. We addressed this issue by introducing ongoing self-reflection on students' learning processes throughout the module and introducing essay questions on exams that targeted cultural reflection on environmental topics.

Based on our experience in carrying out this pilot project, we also have some recommendations for readers considering similar projects. Instructors interested in CBI should explore the Library of Congress website for usable images, recordings, texts, maps, historical documents, and films, as well as a wealth of resources for teachers and researchers. These should be approached with the specific discipline in mind, as well as student level and learning outcomes. We also recommend that instructors familiarize themselves with inquiry-based approaches to optimize their selection and use of environmentally-themed materials in the L2 curriculum. Finally, our recommendation is to motivate students with genuine questions that engage their natural curiosity (and the instructor's) and develop creative ways of investigating those questions.

Conclusion

In the *Charte de l'environnement* unit, students in a third-semester French language class were taught through an experiential learning instructional model that immersed them in the political, social, and cultural dimensions of the environment in French and American contexts. In the pilot and subsequent curricular revisions, we found that the content-based units on environmental sustainability contributed to students' cultural knowledge, language development, and environmental literacy. Combining inquiry, experiential learning, and the study of primary sources, this approach allowed students to pursue more personalized explorations of the target culture and exposed them to research in the language classroom, emphasizing the *process* of research rather than its end product.

In sum, our French language curriculum has increased students' engagement with environmental sustainability. That is, combined with experiential learning on campus, in local communities, and in online exchanges with students in France, student inquiry into the target culture has yielded a more comprehensive outlook on environmental questions, one that surpasses a single standpoint and is informed by historical, literary, cultural, social, and political lenses at once. As Sampon-Nicolas writes, "Clearly, cultural materials and experiential learning activities in combination with literary texts can raise students' preparedness for civic, environmental, and global challenges" (2019, p. 43). Through our continuing efforts to refine our Intermediate French curriculum, we will strive to realize that goal.

Notes

1 *Le Français en (première) ligne* is an exchange network created by Christine Develotte and François Mangenot in 2002, initially between the University of Sydney in Australia and the Université de Franche-Comté in France. It has involved many other universities since and is described on this website: https://www.unicollaboration.org/index.php/2012/06/27/le-francais-en-premiere-ligne-2/.

2 Since 2006, the Library of Congress has awarded Teaching with Primary Sources (TPS) grants to create a nationwide network of organizations providing professional development, curriculum, and projects using the LOC's digitized primary sources. For more information, see https://www.loc.gov/programs/teachers/about-this-program/teaching-with-primary-sources-partner-program/

3 Later, it was moved to week 11 to allow more time to study the text on France's food sustainability efforts and for students' research on campus and in local communities.

References

Assemblée Nationale. (2019). *Les députés en chiffres.* https://www.assemblee-nationale.fr/dyn/vos-deputes

Cammarata, L. (Ed.). (2016). *Content-based foreign language teaching: Curriculum and pedagogy for developing advanced thinking and literacy skills.* Routledge.

Carrico, A. (2019). Reflections on water: Inspiring environmental consciousness through engagement with French texts. In C. A. Melin (Ed.), *Foreign language teaching and the environment: Theory, curricula, institutional structures* (pp. 81–93). New York: Modern Language Association.

Conceição, P. (2020). *The next frontier: Human development and the Anthropocene.* United Nations Development Programme.

Cummins, J., & Sayers, D. (1995). *Brave new schools: Challenging cultural illiteracy through global learning networks.* New York: St. Martin's Press.

Develotte, C., Mangenot, F., & Zourou, K. (2007). Learning to teach online: 'Le français en (première) ligne' Project. In R. O'Dowd (Ed.), *Online intercultural exchange* (pp. 276–280). Clevedon, UK: Multilingual Matters.

Dewey, J. (1897). My pedagogic creed. *The School Journal, 54*(3), 77–80.

Dewey, J. (1963). *Experience and education.* New York: Collier Books.

Discovery Work Group. (2021). *Faculty discovery white paper.* https://discovery.berkeley.edu/faculty-dept-resources/faculty-discovery-white-paper

Freinet, C. (1948). *L'école moderne française, guide pratique pour l'organisation matérielle technique et pédagogique de l'école populaire* (4e éd.). Gap: Ophrys.

Freinet, C. (1994). *Oeuvres pédagogiques.* Paris: Seuil.

Hmelo-Silver, C. E., Duncan, R. G., & Chinn, C. A. (2007, 2007/04/26). Scaffolding and achievement in problem-based and inquiry learning: A response to Kirschner, Sweller, and Clark (2006). *Educational Psychologist, 42*(2), 99–107. https://doi.org/10.1080/00461520701263368

Kolb, D. A. (1984). *Experiential learning: Experience as the source of learning and development.* Englewood Cliffs, NJ: Prentice-Hall.

Krueger, M., & Ryan, F. (Eds.). (1993). *Language and content: Discipline- and content-based approaches to language study.* Lexington, MA: D. C. Heath.

Library of Congress. (2021). *Getting started with primary sources.* https://www.loc.gov/programs/teachers/getting-started-with-primary-sources/

Lightbown, P. M. (2014). *Focus on content-based language teaching.* Oxford: Oxford University Press.

Met, M. (1998). Curriculum decision-making in content-based language teaching. In J. Cenoz & F. Genesee (Eds.), *Beyond bilingualism: Multilingualism and multilingual education* (pp. 35–63). Philadelphia: Multilingual Matters.

MLA Ad Hoc Committee on Foreign Languages. (2007). Foreign languages and higher education: New structures for a changed world. *Profession, 12*, 234–245. http://www.mla.org/flreport

Sampon-Nicolas, A. (2019). Sustainability literacy in French literature and film: From solitary reveries to treks across deserts. In C. A. Melin (Ed.), *Foreign language teaching and the environment: Theory, curricula, institutional structures* (pp. 39–56). New York: Modern Language Association.

Simpson, P. A., & Mueller, M. J. (2019). German is the new green? Language, environmentalism, and cultural competence. In C. A. Melin (Ed.), *Foreign language teaching and the environment: Theory, curricula, institutional structures* (pp. 17–38). New York: Modern Language Association.

Snow, M. A. (2014). Content-based and immersion models of second/foreign language teaching. In M. Celce-Murcia, D. M. Brinton & M. A. Snow (Eds.), *Teaching English as a second or foreign language* (4th ed., pp. 438–454). Boston: Cengage/National Geographic Learning.

Stoller, F. (2002). Promoting the acquisition of knowledge in a content-based course. In J. Crandall & D. Kaufman (Eds.), *Content-based instruction in higher education settings* (pp. 109–123). Alexandria, VA: TESOL.

United Nations Climate Change. (2021). *The Paris agreement.* https://unfccc.int/process-and-meetings/the-paris-agreement/the-paris-agreement

8 Toward Sustainability in German Curricula

Alec Cattell and Belinda Kleinhans

Sustainability, German Studies, and the *Global Readiness Through Language and Culture* Project

Social Sustainability in Foreign Languages

The discussion around sustainability in the curriculum is complex and multifaceted, encompassing environmental protection, economic viability, and social equity. In our educational context, most important are emancipatory models that equip learners with the ability for collective action (Kost et al., 2020). Wals and Jickling (2002) caution against using the term "education for sustainability" because of its prescriptiveness, intellectual determinism, and potential to hinder emerging discourses, especially at the local level (p. 222). Through a comparison of "big brother sustainability" and "grassroots sustainability" (p. 226), the authors build their argument in favor of the latter approach, prioritizing the formation of action-competent citizens who can think critically and work collaboratively to respond to emerging issues, foregrounding these citizens' capacity to "[...] value and respect alternative ways of thinking, valuing and doing" (p. 225).

In seeking to elucidate the role of sustainability in the context of recent curricular innovations in our German language curriculum, which involved the redesigning of several German language courses, we proposed an emancipatory model of sustainability-based curriculum (Wals & Jickling, 2002). This model entails applying democratic and participatory processes to guide course content and learning objectives to link environmental and social concerns (p. 225). This collaborative approach aligns productively with the type of experiential, learner-centered pedagogy of Critical Content-Based Instruction (CCBI) (Sato et al., 2017) and, in particular, Project-Based Language Learning (PBLL) (Beckett & Miller, 2006), the instructional frameworks chosen for the project.

The work presented here was primarily concerned with sociocultural aspects of sustainability in the teaching of German, which provide a solid foundation upon which to foster economic equity and environmental protection. These include themes as poverty, unemployment, and inequality as they relate to economic development and ecological viability (Purvis et al., 2019,

DOI: 10.4324/9781003080183-11

p. 682). Pointing out that economic and environmental concerns have frequently overshadowed the importance of social concerns in the discourse on sustainability, Zygmunt (2016) makes a compelling case that language education is an indispensable component of sociocultural sustainability, serving as "[…] a cornerstone for sustainable development viewed in a holistic way" (p. 116). According to Zygmunt, the role of the educator is to help learners develop the linguistic and (inter)cultural competence they need to be effective leaders in the pursuit of sustainable development. Learning languages and cultures provides opportunities to encounter and engage across human differences, equips individuals to participate in local and international debates, and prepares them to collaborate with others to address issues such as "[…] unemployment, economic growth, pollution, and environmental protection […] ethnicity, tolerance and respect as well as equal opportunities" (p. 122). In the words of Phipps and Gonzalez (2004, p. xv), "languages are a social justice issue."

Sustainability in German Studies

The reconceptualization of the German curriculum at Texas Tech University (TTU) was influenced by the well-known recommendations of the Modern Language Association report 'Foreign Languages and Higher Education: New Structures for a Changed World' (2007). It was also informed by the three fundamental areas of educational practice for reform that Ryshina-Pankova and Byrnes (2017) identified a decade later: (a) a programmatic mission statement anchored in humanistic learning at the collegiate level, (b) a stimulating content- and language-integrated curriculum, and (c) language- and content learning-oriented assessment that can also provide evidence of student learning (p. 425). Within the past decade, several noteworthy innovations in German curricula have responded to the need for content and language integration, from the inclusion of a unit on 'green' business for incorporating critical reflection on environmentally conscious practices in a business German course (Gonglewski & Helm, 2013) to the use of paintings, posters, and films in an integrated language-and-content course to support the development of critical visual and environmental literacies (Ryshina-Pankova, 2013). Some departments have even developed dual degree programs that merge "skill-based, professional training with critical thinking embedded in a full German Studies B.A. [to equip students with] both international career assets and a personal value system that includes a working understanding of social justice and intercultural ethics" (Finger & Tracksdorf, 2019, p. 204). Still others have initiated discipline-wide conversations about what it means to teach German Studies in a global context at all levels of the curriculum, such as the American Association of Teachers of German (AATG)-sponsored seminar at the 2018 German Studies Association and the special issue of *Die Unterrichtspraxis/ Teaching German* facilitated by Coleman and Hock (2019). Simpson and Mueller (2019) approach teaching environmental literacy in the German curriculum from the perspective of "systems thinking" and a holistic approach to ethics in order to encourage their students "to see beyond their immediate needs [...] and to situate human subjectivity in an amplified perspective that emphasizes connections rather than divisions" (p. 19). Simpson and Mueller provide helpful suggestions for teaching environmental literacy through activities, content, and texts, demonstrating that productive connections can be made between foreign language study and disciplines beyond the humanities. Our initiative builds on these insights to conceptualize a transformed curriculum grounded in CCBI and PBLL, as outlined below.

The Global Readiness through Language and Culture Project

This redesign of the German curriculum was possible through participation in an interdisciplinary project entitled Global Readiness Through Language and Culture (GRLC).[1] German undergraduate courses were reconceptualized to raise students' awareness of and engagement with global challenges. Funded by the Texas Tech Center for Global Communication from 2019 to 2021, the GRLC project is designed to empower faculty to implement strategic shifts in their programs to address sustainability both as the content of instruction and as the goal of language education. It involves over 30 faculty members and graduate students working in multiple disciplines housed in the Department of Classical & Modern Languages & Literatures and engages over 3,000 undergraduate students taking courses in the department. All project contributors were asked to include on their syllabus a common statement regarding the goals of the project:

> The Global Readiness Through Language and Culture project aims to prepare undergraduate students studying world languages to (1) identify global issues connected to their major, (2) conduct research on those issues, and (3) demonstrate competencies to address them. The GRLC project showcases the competencies students develop via the study of languages and cultures as highly transferable and crucial for navigating global challenges related to population, resources, technology, information, economics, conflict, and governance.

GRLC was a response to the various changes that foreign languages in higher education had experienced and the challenges that came with those changes.

Institutional, Generational, and Disciplinary Challenges

Enrollments in language programs are steadily declining across the United States (Looney & Lusin, 2019). Overall, fewer students choose to major in German, which means that many miss the opportunity to develop the coveted transferable skills that the study of language, culture, and literature can deliver (Moeller & Abbott, 2018). This indicates the institutional and political constraints placed on the study of languages and prevailing myths about the value of foreign language study in the United States (Gikandi, 2019). Compounding these myths, Sato et al. (2017) point out how neoliberal discourse in higher education has created an unhealthy tension between educational missions and socioeconomic reality, so "[...] not only are language programs pressured to compete against each other in student recruitment and maintenance of the number of enrollments and majors, but they are also forced to prove how productive, efficient, and profitable their programs are, both to university administrators and the general public" (p. 55). Even as institutional support for foreign language study wanes, the term 'global' appears in university mission statements and strategic plans. Coleman and Hock (2019) acknowledge that such institutional commitments are rarely coupled with institutional support for language programs. However, they point out that if we take ourselves out of the conversation, it will "readily move forward without us if we do not insert it into our teaching, scholarship, and programs" (p. 124). Successful initiatives have been documented of interdisciplinary and professional networks to reconceptualize programs. For instance, Finger and Tracksdorf (2019) observed successes in their programs by employing project-based and task-based learning across the curriculum (p. 204) and making connections to other academic areas as well as students' interests and goals (p. 206).

The second challenge is a generational one, as Generation Z has increased demand for practical college degrees. The perception that humanities degrees are not as valuable as professionally oriented majors is especially harmful to language programs; it is difficult to convince GenZ'ers to choose a foreign language major over others they perceive to be more practical (Selingo, 2018, p. 25). While generational shifts like these threaten language programs, acknowledging them is the first step toward turning them into opportunities. For Generation Z, "learning outcomes like critical thinking and analytical reasoning have become paramount, perhaps more vital than subject knowledge, and curricula need to evolve accordingly" (Selingo, 2018, p. 25). The needs and preferences of GenZ align in important ways with principles put forward regarding sustainability pedagogy in higher education: They favor hands-on, experiential learning and opportunities to apply new concepts through project-based learning, undergraduate research, and internships.

The third challenge is a disciplinary one, and it involves ongoing debates among German scholars about the nature, scope, and aims of German Studies. As scholars and educators in this field, we are concerned about ethnonational messaging in German teaching that perpetuates monolithic notions of cultural identity. The notion that the study of German is ideologically neutral is problematic, as it denies the legacy of racism and oppression inherent in the canon, in history, and the present moment. Criser and Knott (2019) point out the need to decolonize the curriculum and describe several manifestations of decolonizing pedagogies that move beyond the inclusion of a single unit or course that engages with diversity issues.

The GRLC project believes that participation in the institutional conversation is essential for the continued existence of language programs and their ability to make meaningful contributions to sustainable development. German Studies can contribute constructively to the challenges of our time by establishing a raison d'etre that resonates with educational institutions and a diverse student body, and the global community. The GRLC project aims to communicate the relevance and value of a language degree to students, parents, university administrators, and colleagues by rethinking curricular content and pedagogies to support social justice and equity.

In the remainder of the chapter, we present the impact of the GRLC project on three courses positioned strategically throughout the curriculum—GERM 2302: *A Second Course in German II*, GERM 3301: *Cultures of the German-Speaking World*, and GERM 4303: *German Classics*. First, we discuss the pedagogical approach and how CCBI and PBLL inform curriculum design and assessment of student learning outcomes; then we describe the scope, objectives, and goals; and finally, we discuss some of the data collected and the students' perceptions of their learning.

Fostering Sustainability through Critical Content-Based Instruction (CCBI) and Project-Based Language Learning (PBLL)

Particularly helpful for the early stages of this curricular project was Wals and Jickling's (2002) concept of "didactical re-orientation" (p. 228)—in the context of integrating sustainability—that "pre-suppose a problem orientation, experiential learning, and lifelong learning" (p. 229). This pedagogy changes what goes on in the classrooms

> from consumptive learning to discovery learning and creative problem solving, from teacher-centered to learner-centered arrangements, from theory-dominated learning to praxis-oriented learning, from sheer knowledge accumulation to problematic issue orientation, and from emphasizing only cognitive objectives to also emphasizing affective and skill-related objectives.
>
> (p. 229)

Many of the new, post-communicative approaches to foreign language instruction, all of them under the content-based framework, share several, if not all, of these characteristics. PBLL is a relatively new pedagogical approach in language curricula. PBLL is "a pedagogical approach focused on engaging students in exploring meaningful, real-world issues and having them seek solutions to intellectually challenging problems that have real-world consequence" (Ferry, this volume) and employs methods that emphasize "critical reasoning, a social (rather than artificial) context for learning, and a focus on the problem instead of information for information's sake or rote memorization" (Ferry,

Table 8.1 Stoller's principles of project-based learning

Principle (P)
1   process and a product orientation
2   co-defined by students to encourage ownership and investment
3   extend over a longer period
4   natural integration of skills
5   dual commitment to language and content learning
6   students work in groups and on their own
7   students take some responsibility for their learning by working with target
8   language resources
9   new roles and responsibilities for teachers and learners
10 tangible final product
students to reflect on both the process and the product

this volume). Stoller (2006) identifies ten central elements of project-based learning, which served as the basis of our instructional approach (see Table 8.1) (Stoller, 2006, p. 24).

Our work draws on several recent studies that indicate multiple benefits of PBLL. For example, Assaf (2018) and Aghayani and Hajmohammadi (2019) have examined the benefits of PBL in English as a Foreign Language (EFL) settings, including boosting learners' motivation and developing their writing skills, digital literacies, and organization, planning, and collaborative skills. Dooly and Sadler's (2016) study on technology-enhanced PBLL, also conducted in an EFL setting, illustrates the importance of context, repetition, multimodality, scaffolding, the use of appropriate materials, and iterative opportunities for language input and production in the implementation of project-based approaches (p. 73). Cox and Montgomery's (2019) study on PBLL in the Spanish foreign language classroom found that the integration of projects requiring collaboration, creativity, critical thinking, and technological literacy led to improvements in learners' behavioral, cognitive, and emotional engagement (p. 844). Building on these insights, the present study seeks to contribute to the discussion on PBLL in German pedagogy while stimulating further conversations about teaching sustainability in German curricula.

Another area in need of reconceptualization is content-based pedagogy. Sato et al. (2017) argue for a re-orientation around the notion of criticality. CCBI aims to enable reflective critiques of power imbalances in society and support the enactment of changes to emancipate and empower people (p. 51). The authors' proposal constitutes a sociocultural turn that urges practitioners to focus on three areas of linkage: (1) connections between language education and other academic fields; (2) language education and education at large; and (3) language education and society (p. 53). Sato et al. stress that, in an ideal CCBI curriculum, language learners

> ...should be active agents who not only strive to acquire the given linguistic and cultural knowledge but can also adopt critical perspectives

when analyzing and evaluating that knowledge. They should have the flexibility and tolerance to understand and accept different views while constantly pursuing change for the better.

(p. 54)

These qualities are urgently needed among people working on sustainable solutions for people, profits, and the planet. They can participate in debates about unemployment, economic growth, environmental protection, ethnicity, or equal opportunities, as Zygmunt (2016) has argued, when they have a foreign language and culture education. By stepping out of familiar ways of thinking and communicating, language learners encounter new modes of thought and action that shed light on the diversity of human experience. This is most effective when course content and activities broaden students' awareness of global challenges and foster their ability to reframe human thought and action (Kramsch & Whiteside, 2008). It is precisely these kinds of changes that we sought to enact in the German program at TTU.

Toward a Sustainable German Curriculum

We strategically redesigned three German courses: GERM 2302 (a second-year language course), GERM 3301 (a third-year course, required for minors and majors), and GERM 4303 (a fourth-year capstone course for majors, also taken by several minors) by adopting PBLL as pedagogy and revising the course content and design according to CCBI principles. Beyond these three courses, all five German faculty members participated in the GRLC initiative by integrating at least one substantial project in at least one of their courses each semester, thus laying the groundwork for complete integration of PBLL and CCBI throughout the curriculum.

We designed the courses by following Stoller's principles of PBL outlined above (see Table 8.1). These brought changes to the content and the teaching approach. On the content side, students were required to co-define their projects, therefore not working simply to fulfill the requirements of a task but co-shaping the project on all levels, which encourages ownership and investment (in accordance with Stoller's, principle 2, hereafter P2). This aspect also meant for instructors to let go of some of the control in the classroom, thereby opening up new roles and responsibilities for both teacher and learner (P8). Especially in the lower language levels (including GERM 2302), there is still a prevailing instruction model where teaching equates to knowledge transmission, incompatible with CCBI. Requiring a tangible final product (P9) that was also assessed, in addition to foreign language proficiency, ensured that students were dually committed to language and content learning (P5).

PBLL introduced three new elements to our courses: Projects were not limited to a single class period or a homework assignment but built over an extended period of the semester (P3). In foreign language classrooms, students were traditionally assigned homework that lost its validity in the subsequent lesson when the class moved on to a different topic or grammatical item. PBLL helped enact a shift away from perceived 'busy work' that does not inspire student investment. This shift allowed students to work on a project for a longer time and become invested in each step leading up to the presentation of the final product. Requiring students to reflect both on the process of creating a tangible product and on the final product itself (P10) allowed for a deeper and more conscious development of a variety of skills. Students were also encouraged to draw on skills beyond those traditionally trained in a foreign language classroom, fostering the natural integration of various skills (P4).

Second-Year Course

Before its redesign, GERM 2302 focused primarily on vocabulary acquisition, grammar drilling, and second language speaking and writing through weekly vocabulary quizzes, grammar exercises as homework, a traditional oral exam, and an in-class midterm and final exam. The course's primary textbook covered limited cultural themes and lacked sustainability content. Similar to Seija and Parra (this volume), we saw a need to go beyond the textbook in order to engage learners with current sustainability issues faced by German-speaking communities. We therefore shifted the approach to a focus on multiliteracies, a concept that transcends a narrow view of literacy as synonymous with reading and writing to "stress the importance of situating language use within socially complex multimodal contexts" (Warner & Dupuy, 2018, p. 117). This ensured a dual commitment to language and content learning (Stoller P5) and led to a more explicit connection between language study, students' majors, and global challenges through the incorporation of two tangible projects (Stoller P9), a narrated video and a bilingual essay. In these projects, students researched various current issues, including migration, multiculturalism, state surveillance, terrorism, fascism, and education. Students' engagement with these global challenge themes vis-à-vis research articles and German-language films supported the social pillar of sustainability by considering the role of cultural artifacts in raising awareness and provoking critical thought about various facets of social justice, civic responsibility, and individual and collective wellbeing. Furthermore, allowing students to choose their own global challenge theme and link it to their academic interests and career goals encouraged student investment and ownership of their learning (Stoller P2).

Work on these projects was carefully sequenced and scaffolded over 15 weeks (Stoller P3) to ensure that students developed writing, speaking, research, analytical, and presentational skills to present their final projects in the last two weeks of the semester. Through a mixture of in-class and take-home work, students learned how to conduct online research, identify appropriate audio and visual materials following copyright law, and create a compelling narrated video using PowerPoint. Similarly, they learned how to find scholarly texts using university library resources and draw on these sources to identify and discuss the portrayal of global challenges in contemporary German-language films. An important goal for the instructors teaching GERM 2302 was to ensure that the projects allowed for critical engagement with content and the development of level-appropriate language skills in a balanced way. For instance, in the first part of a bilingual essay assignment, students wrote a 250-word plot summary of their selected film to practice narrating in the past tense in German, as well as a 250-word description of the main character(s) to practice using adjectives and relative clauses to describe people in German. In the remaining 500 words, students reflected in English on the social sustainability issue or problem presented in the film they watched and discussed in the scholarly article they read, making connections between global challenges and their negotiation in authentic cultural artifacts. This assignment enabled them to develop and showcase their German language skills in the first half of the essay and their critical and analytical thinking skills in the second half of the essay.

These changes to GERM 2302 radically shifted the learning outcomes from a narrow focus on language proficiency and cultural knowledge to a broad focus on digital, textual, technological, and visual literacy, fostering the natural integration of multiple skills (Stoller P4). Furthermore, the projects encouraged exploring connections between language study and global challenges related to social sustainability. Throughout the semester, class discussions frequently turned to the value of studying German in terms of professional development and showcasing transferable skills via ePortfolios. These activities enabled students to reflect on both the process and products of their learning (Stoller P10) and articulate skills beyond German language proficiency such as attention to detail, analytical reasoning, and critical thinking.

Third-Year Course

As part of the GRLC initiative, GERM 3301, formerly known as *German Culture and Society*, was redesigned as a CCBI course and renamed *Cultures of the German-Speaking World* with the goal of re-orienting the course around notions of cultural plurality and intercultural competence. As discussed above (cf. Zygmunt, 2016), these notions are essential for sustainable development at all levels. Course materials and activities were designed to foster critical engagement with received ideas about 'Germanness' through contemporary readings and videos that focused explicitly on human differences such as age, class, dialect, ethnicity, gender, migration, race, region, religion, political systems, and sexual orientation in the German-speaking world. This shift allowed students to learn about and develop an intercultural being, as defined by Phipps and Gonzales (2004, p. 29), by moving from a static, knowledge-bound perspective on culture to a dynamic and critical perspective on the complexities of cultural realities. Therefore, the redesigned version of GERM 3301 enables students to attend to the complexity of L2 literacy, not merely as the interpretation and production of material texts, but also as lived experience: as practices that manifest across multiple languages, cultural contexts, and social ecologies; as a means of accessing and of developing identities, for example as a speaker of a new language.

(Warner & Michelson, 2018, p. 4)

To create a PBLL environment, students were guided through the creation of three digital stories throughout the 15-week semester (Stoller P1 and P3). These projects challenged students to explore their own cultural identities and examine intercultural encounters through the multimodal medium of digital storytelling, as also explored by Anderson et al. (2018, p. 198). In addition to learning about contemporary intersectional experiences in Germany, Austria, and Switzerland through engagement with videos and texts created for native speakers (Stoller P7), students participated in spoken and written activities throughout the semester that culminated in the creation of three digital stories narrated in German (with English subtitles) to articulate insights to others in compelling, genre-appropriate ways (Stoller P4 and P5). The digital storytelling projects, each spread over three weeks, allowed students to develop not only their German language skills but also "language as a skill," meaning that language is a medium through which "communities of people engage with, make sense of and shape the world" (Phipps & Gonzales, 2004, p. 2). These projects, which were crafted around negotiations of self and others and reflections on intercultural encounters, allowed students significant freedom in selecting themes to explore (Stoller P2) while providing opportunities for peer collaboration and feedback (Stoller P6). Leading up to each due date, classroom activities provided training in various aspects of digital storytelling, such as the gathering of audio/visual materials, scriptwriting, storyboarding, voiceover recording, video editing, and providing peer feedback on aspects such as pacing and economy (Stoller P6 and P8). One class period was dedicated entirely to learning how to use the Creative Commons and Europeana to obtain images and sounds labeled for reuse. At the end of the semester, students were invited to share their work with a broader audience at an undergraduate symposium (Stoller P9).

Furthermore, each student was required to write an essay of 500–750 words in English to accompany their German-language videos, thus providing meta-level commentary on the themes they explored and what they learned about the nature of culture in general and cultures in the German-speaking world in particular. Additionally, students' final essays included reflecting on their growth as German-language users and digital scholars and how they honed their techniques to communicate effectively via this multimodal genre, including writing and revising their scripts, creating voiceovers, and selecting images and sounds (Stoller P10). Students were encouraged to include their videos and essay in their ePortfolios to showcase their development of transferable skills. The digital storytelling projects in GERM 3301 provide iterative opportunities for students to reflect, discuss, and learn about their own and others' identities and experiences within an emancipatory approach to teaching sustainability. This explicit focus on social aspects of sustainable development equips students to attend to nuances when engaging with the complex challenges our communities face while recognizing that, because our understanding of others' experiences will never be complete, our learning must be a lifelong process.

Fourth-Year Course

The original catalog description for GERM 4303 read "Readings in German literature through selected works by Hoffman, Büchner, Keller, Kleist, Storm, and Hauptmann." This description is highly problematic for several reasons: It sets as a learning goal the (at least partial) assimilation of a canon, here exemplified by canonical 19th-century authors, forcing students into a passive position (see Phipps & Gonzales, 2004, p. 71). Furthermore, it perpetuates the monolithic nature of the German canon/curriculum by exclusively requiring readings by white male German authors. The inclusion of solely literary texts as 'Classics' also privileges the written word over other modes of cultural production. To address these limitations, the course was fully reconceptualized as a PBLL and CCBI course through the lens of adaptation theory, which allows students to "adjust, to alter, to make suitable" the literary content discussed in class and treats adaptation both as a process and a product (Hutcheon, 2013, p. 7). This characteristic makes it especially suitable for the goals of PBLL. Using excerpts from Hutcheon's *A Theory of Adaptation* (2013), students were trained in the basics of adaptation theory and, in line with Stoller's second principle, co-developed analytical questions to engage with adaptations of a literary text into a different medium. This step allowed students to grasp how to engage with adaptations and exposed them to adaptations that break with the monolithic image of German culture presented by the original text alone. These questions, then, also formed the basis of subsequent assessments of student adaptations, therefore fostering Stoller's eighth principle to revise teacher and student roles and responsibilities.

To train students in the ability to recognize and reflect on different aspects of sustainability as expressed in artistic adaptations and to increase sustainability literacy, they were exposed throughout the semester to three major contemporary adaptations of classical texts: Eric Corbeyran and Richard Horne's *Die Verwandlung* (2010), a graphic novel adaptation of Franz Kafka's famous story; Bridge Markland's *leonce + lena in the box* (2018),[2] a gender-bending one-woman-playback performance for the "generation pop music" adapted from Georg Büchner's *Leonce und Lena*; and Jan Friedrich's modern adaptation of Frank Wedekind's *Frühlings Erwachen* at the Schauburg München (2018).[3] These adaptations included commentary on economic and environmental sustainability, poverty, unemployment, and social inequality. Confronting students through adaptations with multiple versions of the same story allowed them also to develop critical perspectives for analyzing and evaluating by engaging historical distance and cultural differences between German-speaking countries and the US. This trained students' ability to think of values and truths as depending on a specific cultural, historical, and political angle and trained flexibility and tolerance while preparing students for their own project-based engagement with literary texts.

As part of PBLL, students were challenged to adapt literary texts for US society today into a new medium of their choice and update the message of the original text to speak to a contemporary audience, therefore producing a tangible product (Stoller P9). The texts students adapted included a 19th-century ballad of their choice; Rainer Maria Rilke's poem *Der Panther*; Wilhelm Hauff's *Das kalte Herz*; and Max Frisch's *Biedermann und die Brandstifter*. Students chose to adapt these texts into a variety of media, including tweets, video essays, comics, collages, drawings, and even playlists and connected these (for them dated) literary texts to contemporary societal challenges. The multimodal mini adaptations were the basis of many class discussions discussing the process and then the product (Stoller P1), extended over several weeks in the semester (Stoller P3) and allowed students to develop both perceptive and productive multiliteracy skills. In line with Stoller's second principle, students took ownership of their learning when they identified which elements of an old literary text can still speak to an audience today, even outside of Germany, and which elements require updates—and allowing them any type of alteration or update that they deemed necessary. This also ensured that the study of literature was embedded in their social context, fostering an active process of critical reflection and personal response to texts, rather than a passive accumulation of knowledge about them (Phipps & Gonzales, 2004, p. 86) while combining language and content learning (Stoller P5). Students were given a choice to co-create an adaptation or work alone; however, even students working independently were required to reflect, in teams, on their adaptation (Stoller P6).

To ensure reflection on both process and product (Stoller P10), the adaptation project entailed, as the last step, an individual analytical reflection on their own creative adaptation of a German literary text, which formed the basis of the data gathered with the TTU Center for Global Communication (CGC) rubric. Students analyzed their specific adaptation to explore media specificity in the expression of ideas and explain cultural overlap and differences between the historical text, its period and culture, and the world we live in today. By reflecting on which ideas still resonate with audiences today, and which aspects needed to be changed, and how, students gained a deeper understanding of their own culture and world, identifying current challenges and needs to which they could link their adaptation.

Students' projects demonstrated a deeper understanding of various levels of sustainability: Students tackled the sociocultural pillar of sustainability by addressing topics as varied as animalizing discourse and violence (Kafka's *Die Verwandlung*); the links between the lack of Sex Ed and teen pregnancies as well as teenage suicide and Sex Ed in Texas (Wedekind's *Frühlings Erwachen*), immigration in the United States (Frisch's *Biedermann und die Brandstifter*), and the psychological effects of imprisonment (Rilke's *Der Panther*); others dealt with sustainability relating to the environment and resources by focusing on global warming (Frisch's *Biedermann und die Brandstifter*) and environmental decline (Hauff's *Das kalte Herz*).[4]

Outcomes Assessment

Data for this project was gathered from the three GRLC courses between Fall 2019 and Spring 2020. Outcomes evaluation was based on voluntary responses to an anonymous survey and numerical data gathered through a standardized rubric used to assess student projects. The global communication rubric (see Figure 8.1) measures three primary learning outcomes: (1) Content, organization, and delivery; (2) applying knowledge to contemporary global contexts; and (3) cultural and intercultural awareness and engagement. The rubric was adapted to measure learning outcomes according to four performance levels, from benchmark (lowest) to capstone (highest).

The first measured outcome in the CGC rubric, 'content, organization, delivery,' included genre proficiency, information literacy, clarity of structure, and linguistic proficiency, aligning with the shift from only cognitive objectives to affective and skill-related objectives that are essential for PBLL (see Wals & Jickling, p. 229). In a language classroom aiming at sustainability education, this emphasized that language use is always embedded in a larger structure and that to communicate efficiently, lexical and grammatical accuracy alone is not sufficient. The second measured outcome in the CGC rubric, 'applying knowledge to contemporary global contexts,' included students' ability to identify a sustainability hotspot and connect their chosen

Texas Tech University QEP Global Communication Rubric			
	Outstanding [3]	Satisfactory [2]	Minimal Acceptance [1]	Unsatisfactory [0]
Content, organization and delivery	Content and organization are sophisticated for the genre. Delivery choices effectively communicate content to the target audience.	Content and organization are appropriate for the genre. Delivery choices communicate content to the target audience.	Content or organization is minimal for the genre. Delivery choices do not effectively communicate content to the target audience.	Content, organization or delivery is inappropriate for genre and/or target audience, resulting in very low or no audience communication.
Applying Knowledge to Contemporary Global Contexts	Applies knowledge and skills to implement sophisticated, appropriate and workable solutions to address complex global problems using interdisciplinary perspectives independently or with others.	Plans and evaluates more complex solutions to global challenges that are appropriate to their contexts using multiple disciplinary cultural, historical and/or scientific perspectives.	Formulates practical yet elementary solutions to global challenges that use minimal disciplinary perspectives (e.g. cultural, historical and scientific).	Does not define global challenges in a basic way and does not provide varying perspectives and solutions.
Cultural and intercultural awareness and engagement	Articulates sophisticated insights into own and other cultures, and makes connections between own worldview and civic engagement.	Compares and contrasts own culture with other cultures, and identifies appropriate connections between own worldview and civic engagement.	Recognizes minimal components of one's own and other cultures and understands what civic engagement means.	Does not show understanding of own or other cultures, and is not able to describe civic engagement.

Figure 8.1 Global communication rubric. Courtesy of Library of Congress. Using primary sources. Primary Source Analysis Tool.

topic to it by drawing on the subject-specific knowledge and skills acquired in their German class. Finally, the third measured outcome in the CGC rubric, 'cultural and intercultural awareness and engagement,' was used to measure whether students had a sophisticated grasp on how diverse cultures approach their chosen issue of sustainability and whether they were able to establish connections between their worldview and civic engagement.

Three German instructors, two of whom adapted the rubric and a third who received training in the application of the rubric, evaluated the following four courses—Spring 2019: GERM 3301 (15 students); Fall 2019: GERM 4303 (12 students), GERM 2302 (20 students); Spring 2020: GERM 2302 (22 students). The results appear in Table 8.2.

These learning outcomes placed more than 80% of students taking these courses into the upper achievement area (capstone or milestone 2). When discussing the results, we will treat GERM 2302 in Spring 2020 as an anomaly due to the context: The lower achievement in the first learning outcome (content, organization, and delivery) is likely a result of the negative impact of COVID-19 on the ability to train students in a variety of skills and literacies, therefore lowering their score. TTU moved all courses in Spring 2020 to a fully online modality midway through the semester, which interrupted the training for several project-based tasks.

It is apparent that students at the 3,000-level struggled more than their 2,000- and 4,000-level peers in achieving the capstone. Two factors may account for this result: First, this was the first CCBI and PBLL course for the students and the instructor, and unfamiliarity with the model might have impacted the students' ability to perform at the capstone level. Second, the 3,000-level courses at TTU are the first fully content-based courses students take, in which explicit grammar instruction is secondary to content instruction. In 3,000-level courses, disparities between students' language abilities

Table 8.2 Learning outcomes for all four courses

	Course	Semester	Capst.	Milest. 2	Milest. 1	Benchm.
Content,	GERM 3301	Spring 19	40%	46.$\overline{6}$%	6.$\overline{6}$%	6.$\overline{6}$%
organization, and	GERM 2302	Fall 19	80%	15%	5%	0%
delivery	GERM 4303	Fall 19	83.$\overline{3}$%	16.$\overline{6}$%	0%	0%
	GERM 2302	Spring 20	31.$\overline{81}$%	40.$\overline{90}$%	27.$\overline{27}$%	0%
Applying	GERM 3301	Spring 19	26.$\overline{6}$%	60%	6.$\overline{6}$%	6.$\overline{6}$%
knowledge to	GERM 2302	Fall 19	65%	20%	15%	0%
contemp. global	GERM 4303	Fall 19	66.$\overline{6}$%	16.$\overline{6}$%	16.$\overline{6}$%	0%
contexts	GERM 2302	Spring 20	63.$\overline{63}$%	31.$\overline{81}$%	4.$\overline{54}$%	0%
Cultural and	GERM 3301	Spring 19	40%	46.$\overline{6}$%	6.$\overline{6}$%	6.$\overline{6}$%
intercultural	GERM 2302	Fall 19	90%	5%	5%	0%
awareness and	GERM 4303	Fall 19	75%	8.$\overline{3}$%	16.$\overline{6}$%	0%
engagement	GERM 2302	Spring 20	68.$\overline{18}$%	31.$\overline{81}$%	0%	0%

often become apparent for the first time; some students excel in this content-based context, while others struggle to catch up.

Overall, the numerical results demonstrate a higher level of achievement at the 'content, organization, and delivery' level than at 'applying knowledge to contemporary global contexts' (except for GERM 2302 in Spring 2020). We accredit this to the fact that for GenZ, it is easier to relate to the acquisition of concrete, transferable skills, and that allowing students to capitalize on a multitude of skills in PBLL (Stoller's P4) instead of a sole focus on linguistic accuracy leads to better learning outcomes. It also indicates that it is harder for students to connect specific skills associated with language learning and German Studies to issues about sustainability in global contexts. The difference between the second outcome ('applying knowledge to contemporary global contexts') and the third outcome ('cultural and intercultural awareness and engagement') is that the third emphasizes an awareness of intercultural differences and personal, civic engagement, in which the students scored nearly as high as in the first, more skill-based outcome. As the lowest results in the second outcome demonstrate, German Studies and German language education still need to develop strategies to ensure that students can connect skills specific to their German classrooms to significant global sustainability issues.

Conclusions

For practitioners of German, this project provides three essential insights: First, PBLL that trains and assesses a variety of skills beyond language proficiency increases student success and engagement by making the acquisition of transferable skills transparent. In an anonymous survey (which, due to the COVID-19 crisis, only five students were able to complete), 100% of the respondents felt that their German improved while acquiring additional transferable skills traditionally not associated with a German degree, such as video editing, adaptation, time management, critical and creative thinking, analytical skills, digital literacy, visual communication, and awareness of global challenges. An additional benefit to implementing PBLL and CCBI was that students acquired multiliteracies that equipped them with additional transferable skills and boosted problem-solving, demonstrating that PBLL and CCBI are valuable tools not only for sustainability education but also for sustainable education.

Second, even if instructors are restricted to teaching the literary canon, PBLL (especially through digital storytelling and student adaptations) offers the means to shift CBI to CCBI by ensuring a "learning process and content sensitive to the ideas, values, interests, and concepts embodied by the learners themselves" (Wals & Jickling, 2002, p. 225). The inclusion of global challenges in sustainability that can, thanks to the foreign language context, be looked at from different angles, fosters critical thought. Finally, especially in the lower to intermediate-level courses, multimodal projects provide students the opportunity to express ideas for which they do not yet have the linguistic abilities in German, therefore mitigating the sense of frustration that students often experience in courses that focus exclusively on linguistic accuracy and fluency.

Notes

1 For more information on the GRLC project, see https://www.depts.ttu.edu/classic_modern/globalreadiness/index.php.
2 To get an impression of this piece, see https://www.bridge-markland.com/index.php/repertoire/leonce-lena-in-the-box.
3 To get an impression of this adaptation, see https://www.schauburg.net/de/fruehlings-erwachen.
4 To see an example of student creative work, see the short film adaptation of Hauff's *Das kalte Herz*: https://www.youtube.com/watch?v=zci77h3SeTE.

References

Aghayani, B., & Hajmohammadi, E. (2019). Project-based learning: Promoting EFL learners' writing skills. *LLT Journal: A Journal on Language and Language Teaching, 22*(1), 78–85. https://doi.org/10.24071/llt.2019.220108

Anderson, J., Chung, Y. C., & Macleroy, V. (2018). Creative and critical approaches to language learning and digital technology: Findings from a multilingual digital storytelling project. *Language and Education, 32*(3), 195–211. https://doi.org/10.1080/09500782.2018.1430151

Assaf, D. (2018). Motivating language learners during times of crisis through project-based learning: Filming activities at the Arab International University (AIU). *Theory and Practice in Language Studies, 8*(12), 1649–1657. http://doi.org/10.17507/tpls.0812.10

Beckett, G. H., & Miller, P. C. (Eds.). (2006). *Project-based second and foreign language education. Past, present, and future.* Charlotte, NC: Information Age Publishing.

Coleman, N., & Hock, L. (2019). Introduction: German studies go global. *Die Unterrichtspraxis/Teaching German, 52*(2), 124–129. https://doi-org.lib-e2.lib.ttu.edu/10.1111/tger.12094

Corbeyran, E., & Horne, R. (2010). *Die Verwandlung von Franz Kafka.* München: Knesebeck GmbH & Co. Verlag.

Cox, C. B., & Montgomery, C. (2019). A study of 21st-century skills and engagement in a university Spanish foreign language classroom. *Foreign Language Annals, 52*(4), 822–849. http://doi.org/10.1111/flan.12426

Criser, R., & Knott, S. (2019). Decolonizing the curriculum. *Die Unterrichtspraxis/Teaching German, 52*(2), 151–160. https://doi-org.lib-e2.lib.ttu.edu/10.1111/tger.12098

Dooly, M., & Sadler, R. (2016). Becoming little scientists: Technologically-enhanced, project-based language learning. *Language Learning & Technology, 20*(1), 54–78. http://doi.org/10125/44446

Ferry, M. M. (2021). Project-based language learning in the classroom. In M. J. de la Fuente (Ed.), *Sustainability-based foreign language teaching and learning* (pp. 50–64). London: Routledge.

Finger, A., & Tracksdorf, N. (2019). Go global, build networks, create nodes: Integrating the humanities and the professions. *Die Unterrichtspraxis/Teaching German, 52*(2), 203–212. https://doi-org.lib-e2.lib.ttu.edu/10.1111/tger.12105

Friedrich, J. (Director). (2018). *Frühlingserwachen. Eine Kindertragödie von Frank Wede-kind*. [Theatre performance]. Schauburg München.

Gikandi, S. E. (2019). President's column: Language matters. *MLA News-letter, 51*(2), 2–3. Retrieved from https://www.mla.org/content/download/110333/2408099/NL_51-2_web.pdf

Gonglewski, M., & Helm, A. (2013). LaissezFair: A case for greening the business German curriculum. *Die Unterrichtspraxis / Teaching German, 46*(2), 200–214. http://doi.org/10.1111/tger.10141

Hutcheon, L., with O'Flynn, S. (2013). *A theory of adaptation* (2nd ed.). London and New York: Routledge.

Kost, K., Nolan, D., & Peabody, S. (2020). Sustainability and German studies: From ecocriticism to community engagement. *German Studies Review, 43*(2), 377–394. http://doi.org/10.1353/gsr.2020.0048

Kramsch, C., & Whiteside, A. (2008). Language ecology in multilingual settings: Towards a theory of symbolic competence. *Applied Linguistics 29*(4), 645–671. https://doi-org.lib-e2.lib.ttu.edu/10.1093/applin/amn022

Looney, D., & Lusin, N. (2019). *Enrollments in languages other than English in United States institutions of higher education, summer 2016 and fall 2016: Final report*. Modern Language Association. https://www.mla.org/content/download/110154/2406932/2016-Enrollments-Final-Report.pdf

Markland, B. (Creator and performer). (2018). *leonce + lena in the box*. [Theater Performance]. Tony – Die Agentur.

Moeller, A. J., & Abbott, M. G. (2018). Creating a new normal: Language education for all. *Foreign Language Annals, 51*, 12–23. http://doi.org/10.1111/flan.12310

Phipps, A., & Gonzales, M. (2004). *Modern languages: Learning and teaching in an intercultural field*. London: Sage.

Purvis, B., Mao, Y., & Robinson, D. (2019). Three pillars of sustainability: In search of conceptual origins. *Sustainability Science, 14*, 681–695. http://doi.org/10.1007/s11625-018-0627-5

Ryshina-Pankova, M. (2013). Understanding "greening Germany" through images and film: A critical literacy approach. *Die Unterrichtspraxis / Teaching German, 46*(2), 163–184. https://www.jstor.org/stable/unteteacgerm.46.2.163?seq=1

Ryshina-Pankova, M., & Byrnes, H. (2017). Embracing the language-educational challenge of FL departments: Reflections on ways forward. *The Modern Language Journal, 101*(2), 424–427. https://doi-org.lib-e2.lib.ttu.edu/10.1111/modl.12420

Sato, S., Hasegawa, A., Kumagai, Y., & Kamiyoshi, U. (2017). Content-based instruction (CBI) for the social future: A recommendation for critical content-based language instruction (CCBI). *L2 Journal, 9*(3), 50–69. https://doi.org/10.5070/L29334164

Seija, J. M., & Parra, M. L. (2021). Engaging students with social, cultural, and environmental sustainability topics in the Spanish-speaking world: A re-imagined beginning Spanish curriculum. In M. J. de la Fuente (Ed.), *Sustainability-based foreign language teaching and learning* (pp. 87–104). London: Routledge.

Selingo, J. J. (2018). *The new generation of students: How colleges can recruit, teach, and serve Gen Z*. Washington, DC: The Chronicle of Higher Education.

Simpson, P. A., & Mueller, M. J. (2019). German is the new green? Language, environmentalism, and cultural competence. In C. Melin (Ed.), *Foreign language teaching and the environment: Theory, curricula, institutional structures* (pp. 14–25). New York: The Modern Language Association of America.

Stoller, F. (2006). Establishing a theoretical foundation for project-based learning in second and foreign language contexts. In G. H. Beckett & P. C. Miller (Eds.), *Project-based second and foreign language education* (pp. 19–40). Greenwich, CT: Information Age.

Wals, A., & Jickling, B. (2002). "Sustainability" in higher education: From doublethink and newspeak to critical thinking and meaningful learning. *International Journal of Sustainability in Higher Education, 3*(3), 221–232. https://doi.org/10.1108/14676370210434688

Warner, C., & Dupuy, B. (2018). Moving toward multiliteracies in foreign language teaching: Past and present perspectives … and beyond. *Foreign Language Annals, 51*(1), 116–128. https://doi.org/10.1111/flan.12316

Warner, C., & Michelson, K. (2018). Introduction to the special issue: Living literacies: L2 learning, textuality, and social life. *L2 Journal, 10*(2), 3–15. https://doi.org/10.5070/L210238971

Zygmunt, T. (2016). Language education for sustainable development. *Discourse and Communication for Sustainable Education, 7*(1), 112–124. https://doi.org/10.1515/dcse-2016-0008

Part 3

Exploring Interdisciplinary Collaborations toward Sustainability Education in FL Programs

9 Cross-Disciplinarity at the Core

Teaching Sustainability in a Business German Course

Margaret Gonglewski and Anna Helm

Introduction

Germany is often seen as a sustainability rock star, and with good reason. The country has successfully initiated a total 'green' economic transformation while maintaining its standing as the world's fourth largest economy. Through its *Energiewende* (energy transition) Germany has charted a course for eliminating nuclear power and coal, and has committed—through national law—to the Paris Accord carbon neutrality target by 2050 (Climate and Energy, n.d.). It has heavily invested in green technologies and jobs: By mid-2020, nearly half of energy sector workers were in the field of renewable energies (Climate and Energy, n.d.). This success stems not only from economic policies and environmental law, but also from citizen activism that played a crucial role in getting the energy transition accepted by the majority of Germans, by creating 'citizen energy cooperatives,' which place ownership of renewable energy production into the hands of communities instead of large corporations (Wieg, 2013). The serious grassroots-level commitment to a sustainable future extends to Germany's youngest generation, among the most active in the worldwide 'Fridays for Future' movement of school-age children staging weekly demonstrations demanding climate justice and equity for all (Fridays for Future, 2020).

Given Germany's reputation as a green pioneer, it is no surprise that German instructors and language programs are developing courses focused on sustainability. Some have created language courses on 'Green Germany,' addressing a broad array of topics ranging from recycling to climate change (Becker et al., 2013; Ryshina-Pankova, 2013); others are reconceptualizing the German language curriculum toward sustainability (Cattell & Keinhans, this volume); others have presented more narrowly focused units, e.g., on the phase-out of nuclear power (Berg, 2013). These examples demonstrate that sustainability welcomes—if not demands—a cross-disciplinary approach, one that broadens students' understanding of sustainability-related content while ensuring the applicability of their language and cultural competence. Such an approach would find its roots in the 'Language Across the Curriculum' movement, which marries language learning directly to other disciplines, thereby demonstrating the applicability of languages in the broader curriculum and

DOI: 10.4324/9781003080183-13

in learners' other chosen fields—including those not traditionally viewed as connectable (Bettencourt, 2011). In spite of this call for cross-disciplinarity in teaching sustainability, reports on actual collaboration between German language faculty and colleagues from other disciplines are rare. Without the opportunity for direct collaboration, German language faculty typically rely on their own knowledge, skills, and interests to guide them in their course development and delivery.

This paper presents an approach for teaching sustainability in the inherently cross-disciplinary context of a business language course. The approach's pedagogical framework is content-based instruction (CBI), undergirded by the notion that the content of sustainable business is the "vehicle" through which the learning of the second language can best occur (Brinton et al., 2003, p. 182). We present a unit focused on Germany's *Energiewende,* which integrates content and pedagogical tools from the disciplinary field of international business. At the unit's center is a business case featuring a citizen energy cooperative—residents banding together to invest jointly in local renewable energy projects. The authors' differing disciplinary backgrounds bring a unique dimension by intertwining perspectives from business, a discipline that anchors its analysis in data and frameworks, and German studies, which promotes deep analysis from a cultural perspective. In the sections that follow, we first make the case for using CBI in a Business German course focused on sustainability. Then, we provide a detailed description of the unit, the materials and tasks that serve as the medium and means for content and language learning. We conclude with an outcomes analysis based on input from students' self-assessment of their achievement of the unit's learning objectives and perceptions of their experience with the unit.

Framework for the Sustainability Unit

Language for Specific Purposes as Content-Based Instruction

Content-based language instruction has been defined as "the integration of particular content with language-teaching aims" (Brinton et al., 2003, p. 2). Using this approach, language teachers "are asked to let the content dictate the selection and sequence of language items to be taught rather than vice versa" (Brinton et al., 2003, p. 2). The advantages of centering content are many: It builds on learners' previous knowledge and potentially stimulates learner motivation as it showcases the language's immediate practical use (Lafford, 2012). While language forms are not meant to fall by the wayside in CBI, they are also not guiding what is being taught and how, which, in the past, led to a "decontextualized, rule-governed, and 'scientized' understanding of language" (Byrnes, 2005, p. 284).

According to Brinton et al. (2003) "language for specific purposes (LSP) is perhaps the best known and most documented of the content-based language models" (p. 6). The content of any LSP course is taken from the professional field serving as the focus area and drives the language being used and taught in the course. Byrnes (2005) notes that for LSP, at the forefront stands the exploration of "disciplinary ways of knowing," expressed and transmitted through the language, e.g., through contexts and genres typical to a field (p. 289). Since each discipline employs what Byrnes calls "discursive and lexico-grammatical patterns of knowing" (2005, p. 289) particular to that discipline, students should learn those language patterns and practices to prepare for operating appropriately within that field. This is especially applicable at the advanced levels of language learning, which Byrnes identifies as "traditionally the most obvious place for the linking of language and content" (2005, p. 289). With cross-disciplinarity at its core, LSP addresses the urgent need expressed in the Modern Language Association report (2007) for language educators to branch out beyond traditional content to prepare students for their engage-ment within the 'changed world' they will enter after graduating. LSP courses immerse students in content areas beyond those commonly offered in lan-guage programs, providing different "sources of cultural information" as well as different genres and communicative contexts (Long, 2017, p. 3). With its in-tegration of language learning within field-specific content from professional domains, LSP fosters "*Transferable Skills for the 21st Century,*" comprising critical thinking, adaptability, intercultural competence, and collaboration (King de Ramirez & Lafford, 2018).

Business and Language: Mutual Calls for Collaboration

Among the diverse subfields of LSP, business language is the most common in terms of course offerings and faculty research interest (Doyle, 2012; Long & Uscinski, 2012; Sanchez-Lopez et al., 2017). Yet many faculty shy away from business language courses because of the challenges, namely a lack of content knowledge, materials, support, or opportunity to work across disciplinary lines (Grosse & Voght, 2012). Byrnes (2013) points out that a "particularly plausible route for making interdisciplinary connections and attaining a recognized programmatic status and more stable administrative support is via CIBERs" (p. 3). Through US Department of Education funding, these Centers for In-ternational Business Education and Research (CIBERs) have enabled scores of business language teaching and research initiatives since 1988. Indeed, our university's CIBER has opened the doors between our once siloed Business School and College of Arts and Sciences, which has led to numerous business language projects (see, e.g., Gonglewski & Helm, 2010, 2013, 2014, 2017).

Business and language collaboration has been further facilitated by appeals within the field of business to embrace liberal arts learning. Parallel to the MLA call to better prepare language students for the challenges in the new globalized context is the Carnegie Report's call to reform business education in "preparation for twenty-first century life" (Colby et al., 2011, p. 49) through skills gained from a liberal arts education. The report recognized that educating students for a complex and demanding future requires skills ranging from those typically taught in business schools (data analytics, finance, etc.) to those taught in the liberal arts (deep research, critical reflection, etc.) that ultimately must synergize to develop future leaders who can think analytically, critically, and creatively, to benefit their organizations and society as a whole. The authors emphasized reciprocity, as mutual outreach strengthens the ability of business and liberal arts to "animate the educational potential" of both (Colby et al., 2011, p. xi).

Sustainability as Content Focus in Business German

Beyond the urgency surrounding the topic as a global priority, sustainability provides us with a meeting point where we can contribute equally from our disciplinary backgrounds, experience, and interests. Melin (2019) underscored this asset of sustainability, noting that through "cultivating connections with other disciplines where this content is found, foreign language programs can construct their identities anew" (p. 2). While Melin was referring to building bridges "between the humanities and the sciences," we note that sustainability is eminently suited for treatment by many different disciplines, as creating a sustainable future requires solutions that leverage knowledge and approaches from multiple perspectives. Ducate (2016) characterized sustainability as being "influenced by social issues, economics, ethics, and culture," requiring "insights from other fields, including philosophy, psychology, history, and literature to stimulate change" (p. 69). Business Language Studies is especially well positioned to address sustainability content, as it provides a multidimensional approach through its already extant interdisciplinarity (Doyle, 2012).

Many see German and sustainability as a natural combination within German language/culture programs. Ducate (2016) called the theme-duo a "logical draw for students around the university" since it capitalizes on Germany's reputation as playing an "innovative and leading role in the field" and on students' growing interest in sustainability (p. 70). To Berg (2013), Germany is already viewed as a green champion by Americans who have experienced the country's environmental "intentionality," whether through individual actions, such as recycling, or through policy, like conservation efforts (p. 215). Equally compelling is to examine how "[p]olitical activism and cultural history have shaped environmentalism in Germany" (Simpson & Mueller, 2019, p. 19), particularly in the context of high-risk economic decisions by the government and businesses.

As content in a Business German course, Germany's approach to sustainability, particularly in the area of energy, provides a rich context for meaningful language learning. Beyond offering engaging and relevant factual content, it can inspire learners' sense of individual responsibility, encouraging them to see themselves as agents of change, as sustainability professionals themselves must be (Sacco et al., 2014). Simpson and Mueller (2019) argue that environmental literacy should be a "significant learning outcome alongside proficiency in the target language" (p. 19) with the ultimate goal of fostering in students a self-conceptualization as global citizens with a responsibility to the world not limited by national borders. A Business German course using CBI with focus on sustainability themes is, thus, fertile ground for nurturing a critical disposition in learners so that they can "act on and adapt to diverse and changing environments in a sensible and nuanced manner" in order to have a positive impact on the world we all share (Sato et al., 2017, p. 58).

The Business German sustainability unit presented here aimed to capture the potential benefits of a cross-disciplinary collaboration with the additional goal of inspiring students to see their own potential impact on the world around them. Below, we present the unit, situating it first within the course context and then presenting the learning objectives, which include, true to a 'balanced CBI' approach, both content and language goals.

Business German Sustainability Unit

Context and Learning Objectives

The three-week unit was carried out within the context of a fourth-year, CBI Business German course (GER 2111) where students had a demon-strated oral proficiency of at least 'Intermediate High,' as measured by the American Council on the Teaching of Foreign Languages oral proficiency guidelines (ACTFL, 2012). Offered every spring semester, Business German attracts students ranging in class year, however they must have completed the third-year language course (Intermediate German II, GER 2010) or equivalent. Most students who take the course have spent significant time in a German-speaking country, e.g., studying or interning abroad. Expertise or experience in the field of business is not required, but students must have an interest in business-related topics. In Spring 2020, when this curricular intervention was piloted, ten students were enrolled, all of whom fit this general profile.

Using 'theme-based instruction' (Brinton & Snow, 2017), the course was structured around three themes/units presenting topics both interesting and relevant to students. Weigert (2004) advocated this approach for Business German as it allows for selection and compilation of authentic and up-to-date materials as well as level-appropriate task creation, in contrast to traditional textbooks. Theme one focused on applying for professional jobs in Germany, a process strongly steered by cultural norms regarding hierarchies, notions of privacy, etc. Students explored and described their own interests, values, abil-ities, and characteristics in preparation for presenting themselves in a public and professional context. In theme two, students examined Germany as a business location, learning key factors spanning the country's geography and economic system as well as its major industries and companies. By the third theme of sustainability, students were armed with language skills and content knowledge to support their cross-disciplinary work in the unit, which would feature content and tools from international business.

For each unit, content and learning outcomes (LOs) were laid out. In the case of the sustainability unit, LOs related to German culture, sustainability, business, and language development (see Table 9.1). Learners were challenged to understand the historical, political, economic, and social context of Germany's transition from fossil fuels to renewable energy sources, while using language and business tools to grapple with a problem presented in a business case. Because a key goal was to highlight the role of everyday people in the energy transition, working together towards equitable access to investment in—and benefit from—local renewable energy projects, the social aspect of sustainability featured prominently in the LOs.

Table 9.1 Sustainability unit learning objectives

Content	German culture and sustainability	• Define sustainability core concepts (particularly related to *Energiewende*) as used in Germany • Demonstrate a firm grasp of cultural significance of ecologically conscious living in Germany • Compare/contrast sustainability perspectives and practices in Germany and the US • Explain the role of citizen activism in bringing about change in Germany • Present an example of own ecologically conscious actions
	German culture, sustainability, and business	• Use data to analyze current state of renewable energy in Germany (focus on the role of energy cooperatives within the national context of energy production) • Identify drivers of the current competitive landscape in Germany's energy sector (focus on regulatory and political environments) • Apply Hofstede's 'cultural dimensions' as a tool for analyzing receptiveness to renewable energy in Germany and the approach of citizen energy cooperatives • Connect cultural importance of ecologically conscious living and green business in Germany (focus on renewable energy sector)
Language		• Expand vocabulary for understanding and discussing sustainability in German • Apply new knowledge of grammatical tools in German (e.g. *indem*-phrases) for understanding and interpreting text • Use appropriate language to describe, explain, compare/contrast, and interpret information in charts/graphs

Although the LOs are delineated into three categories, there is obvious overlap. As students learn to "define core concepts in the area of sustainability [...] as used in Germany," they quickly learn, for example, that the definition of sustainability has its roots in an 18th-century business decision (see below), giving this goal a crucial business facet. Similarly, to achieve this same LO, students must "expand their vocabulary for understanding and discussing sustainability in German," an explicit language-focused objective. This intertwining of LOs across focus areas mirrors the cross-disciplinarity at the center of our teaching approach.

Case Studies

Working across disciplines meant integrating business content but also using pedagogical approaches from business. Here, the primary vehicle to teach about Germany's energy transition was a *business case*. Cases have a long tradition in US business schools, where they foster a capacity to think critically, creatively, and strategically (Forman & Rymer, 1999). A case's decision-making dilemma involves students in the content, which typically features a story-like narrative and numerous appendices with facts and figures that together construct a realistic business scenario. Also a useful tool for teaching advanced language, cases expose students to multiple perspectives and sometimes contradictory narratives to be understood and synthesized (for a discussion of business case use for language teaching and learning, see Gonglewski & Helm, 2010; Takami, 2008). The unit also utilized Hofstede's framework of 'cultural dimensions' (2011), which offered a way for students to look through the other discipline's lens to analyze cultural differences. Using extensive survey results, Hofstede formulated six dimensions along which cultural values can be analyzed, such as individualism versus collectivism. Our aim in using these business-steeped approaches was to engage students with tools outside of their disciplinary comfort zone, thereby gaining skills to complement and expand their existing language and cultural knowledge.

German Language Development

Finally, the unit featured explicit attention to students' learning of German. The importance of careful planning for second/foreign language (L2) learning in content-based programs has been emphasized, as language will not "emerge incidentally" (Lyster, 2017, p. 111). Weigert (2004) argued for integrating language learning goals specifically at the advanced level, where they are often neglected or purposefully left out due to a presumption that students have already "'mastered' the language in lower division courses" (p. 131). De la Fuente (2019) similarly observed that without explicit focus on L2 development, students' gains can fall short of expectations, as measured both by direct (performance) and indirect (self-evaluation) assessments. In the present project, language components to receive focus were identified based on the prevalence of specific vocabulary and grammatical forms in the selected authentic materials and within the topic of renewable energy. Throughout the unit description, we elucidate how language-focused tasks were integrated while maintaining solid footing within the sustainability- and business-related content.

Description

Introduction to Sustainability and Germany's Energy Transition

The unit began with an activity to draw students into the general topic by asking them to reflect on their behavior vis-à-vis the environment (see Table 9.2). Students first watched a multi-part video called *Was machst du für die Umwelt?* (What are you doing for the environment?) featuring Germans demonstrating their environmentally-friendly actions in everyday life and clarifying the value of personal responsibility for environment and community. The scenes familiarize students with Germany's sustainability culture—extensive recycling system, organic products, bicycle culture—as well as the discourse of environmental activism, providing language (words, wordings) and attitude (emphasis, tone) to emulate in the follow-up task of creating a similar video. That task transformed students from observers to participants and put them 'in conversation with' the Germans in the video, thereby linking sustainability content to what Berg (2013) called learners' "emotional selves" in order to "build their awareness of their own culture as well as gain knowledge of the German cultural perspective" (p. 221). Placing learners in this active and self-reflective role early on aimed to facilitate their recognition of themselves as potential agents of change in the narrower topic of renewable energy.

Before diving into the energy transition, students explored a definition of sustainability from the German perspective, meant to challenge them to see the concept from a different cultural standpoint. They engaged in a direct comparison between the English and German Wikipedia definitions of sustainability. Student teams immediately identified 'superficial' differences such as the length and complexity of the first sentence of the two definitions, which allowed them the opportunity to pull apart a syntactically complex German sentence while considering cultural facets expressed within the genre of Wikipedia definitions. Students examined the contrasting presentations of the term's history, noting that the German entry attributed the concept to Carl von Carlowitz, the 18th-century forester who, in response to the decimation of wood supply, proposed that only as many trees should be cut as could be replaced each year. In contrast, the English entry did not mention Carlowitz. Similarities across the two entries did not escape students: They saw that both featured a three-pillar model of sustainability—new to nearly all students—with economic, social, and environmental facets playing equally important roles.

Table 9.2 Sustainability unit sequence and activities

Day	Topics/materials	Output-based activities
1	- Defining sustainability in the German context - Video: environmentally-friendly actions	- Create/post video response - Read/analyze definitions of sustainability
2	- UN's 17 sustainable development goals - Three-pillar model of sustainability - Sustainable development: Germany's ministry of the economy: SDGs 7, 8, 9, 12	- Watch German video: Defining sustainability - Visit German website on 17 SDGs: Goal 7 - Identify/analyze *indem*-phrase use in text - Build vocabulary (expanding semantic fields), here and throughout
3	- Citizen activism in Germany; the Green Party and energy cooperatives - Energy transition - Language (talking about graphs)	- Read text: Germany's energy transition and energy cooperatives
4	- Business case: Citizen Energy Cooperative Starkenburg	- Reread business case; answer comprehension, analysis questions - Identify/analyze use of *indem*-phrases in text
5	- Lecture: Hofstede's Cultural Dimensions - Hofstede's C. Dim: Germany and US	- Compare/contrast Germany and US using Hofstede framework
6	- Final assignments: Parameters	- Compose new case ending - Apply Hofstede framework: Understanding success of Germany's citizen energy cooperatives

To drive home the revelation that sustainability reaches beyond environmental issues, students were introduced to the UN's 17 Sustainable Development Goals (SDGs), which "address the global challenges we face, including those related to poverty, inequality, climate change, environmental degradation, peace and justice" (United Nations, n.d.). Students examined the German SDG chart, with visual depictions of each goal without textual explanation. They first interpreted the images to determine the actions portrayed and then clicked on the image to check their interpretation. Once familiar with the SDGs, students examined the connection between sustainability goals and economic policy by reading an official statement from Germany's department of economics and energy that expresses the government's commitment to 'Agenda 2030.' The statement reiterated familiar themes to students: It identifies the three pillars of sustainability as guiding principles and highlights SDGs that receive the department's particular focus, notably "7. Affordable and Clean Energy" (BMWi, n.d., p. 2). Diving deeper into goal seven, the class examined the plans for achieving carbon neutrality while continuing Germany's strong global economic leadership role.

In terms of German language development, the use of authentic materials provided the context through which students could actively expand their sustainability-related vocabulary using semantic fields. This encourages students to build out their lexicogrammatical knowledge from one core word and concept to many related ones, whether directly related ("form-based derivational") or linked theme-wise ("content-based associative") (Weigert, 2004, p. 141). In the sustainability unit, the semantic fields were not exclusive to one text but instead were linked to the larger theme that unified all unit materials. The semantic field built around *nachhaltig* (sustainable), for example, featured derivational words such as the noun *Nachhaltigkeit* (sustainability) and the adjectival expression *nachhaltiger Konsum* (sustainable consumption); content-based associative words included phrases like *umweltfreundlich sein* (to be environmentally friendly). Rather than receiving English translations, students recorded use of these terms within unit materials to glean meaning from multiple contexts. With respect to advanced L2 grammar, the unit used an explicit language focus on a grammatical structure used to indicate the means of doing something (by...), the *indem* phrase, prevalent in the unit's texts. In doing so, we followed Lyster's (2017) scaffolded instruction approach of using "pre-planned activities that draw students' attention to language features by means of noticing and awareness activities followed by opportunities to use the features during guided and autonomous practice" (p. 119). First, students' attention was drawn to the boldface *indem*-phrases in the SDG texts; then they analyzed their use both grammatically (how they function) and semantically (their meaning across various contexts). Throughout the remainder of the unit, students were guided in noticing and practicing this form in preparation for the final writing task.

Business Case: Citizen Energy Cooperative

To understand the pivotal role that citizens played in getting the energy transition off the ground, students needed to first grasp the history of grassroots activism in Germany. To this end, the unit allotted a full class period to the notion of sustainability-related citizen initiatives to help students square this type of activism with what they had learned or experienced about German cultural values. As an example, the story of the founding of Germany's Green Party in the 1980s was presented. Equally important to learning that history is understanding that party's role today. Skimming the party's platform on their website, students identified the three sections reflecting the now familiar pillars of sustainability, areas in which the party is active in the legislature. The repetition of this content from the start to the end of the unit helped scaffold students' growing content knowledge about sustainability in German political life, while highlighting citizens' role in affecting change.

Everything in the unit built up to an authentic business case that placed students in the role of protagonists and decision-makers. The case featured the Citizen Energy Cooperative Starkenburg, a group of residents who, in 2010, started pooling their resources to invest in local renewable energy projects, making wind turbines more affordable to people in the region. It was because of citizen activists like this group and their "widespread ownership of renewable energy production" that the German energy transition was successful (Wieg, 2013, p. 4). In 2017, however, new legislation phased out the feed-in tariffs that had made small-scale organizations financially viable and thereby threatened the future of citizen energy cooperatives.

Framed as a short story, with context, characters, and their motivations, the Starkenburg case tells the cooperative's story, their various additional projects (solar, biomass), and the problem they faced (new legislation). Video accompanying the case gives voice to the main characters, with testimonials about citizens' personal and financial investment. Crucial to the business discipline, the case supplies supplementary data in charts and graphs to bolster the narrative with relevant information, e.g., Germany's renewable energy production and use spanning 1991–2019. This presented students the chance to hone their ability to talk about information presented visually, a skill expected within the professional business world. To help them develop that skill in German, students were provided with language-specific information, so that they could comprehend, analyze, and communicate graphic content.

After reading the case, students were introduced to Hofstede's cultural dimensions in German by the business professor and co-author. Students learned how Hofstede's dimensions express themselves in business and other human interactions and why they are relevant to international business operations. After learning the differences between the US and Germany in Hofstede's cultural dimensions, students compared and contrasted their understanding of German and US-American culture, but generally agreed that it corresponded to their cultural experience, particularly regarding long-/short-term orientation and individualism/collectivism. Building on that discussion, they were invited to contemplate the connection between a country's cultural dimensions, according to Hofstede's scale, and its government's and citizens' approach to sustainability. Students again focused on the—perhaps more salient to them—dimensions of long-term orientation and collectivism as indicative of cultural differences expressed in attitudes towards sustainability. They noted Germany's long-term commitment to addressing the sustainability goals in economic policy and action, and the citizens' joint investment in renewable energy to benefit the entire community.

As a culminating task, students completed a multi-part assignment akin to what business students would do following a business case. During class, they debated potential ways to address the problem posed, namely, how to move forward in spite of legal changes. Placing themselves into the protagonist's shoes, they presented their solutions to other coop members (classmates). They then composed a new case conclusion by synthesizing information obtained from online research (the coop's website, news reports) and matching the case genre (like a short story). A separate task required students to apply the Hofstede framework to reflect on why the citizen energy cooperative approach was successful in Germany and to surmise whether it would work in the US context. Here they drew on content knowledge gained in the unit and incorporated it into their analysis while demonstrating their ability to put new language skills to use.

Evaluating Learning Outcomes

At the end of the sustainability unit, students completed a survey and provided written reflections on their learning experience. De la Fuente (2019) notes the value of "self-evaluation tools" for obtaining valuable information about students' learning vis-à-vis the stated course learning goals (p. 141). Thus, the survey contained a list of statements based on the unit's learning objectives (see Table 9.1) such as, "After working with the sustainability unit, I am now able to define core concepts in the area of sustainability [...]." Using this indirect assessment measure, we could begin to understand the effectiveness of the unit and the cross-disciplinary content-based approach. In the survey, students rated their achievement of unit LOs by selecting a level of agreement on a five-point Likert scale.

Not surprisingly, several general comments noted students' frustration over the adjustment to the online environment due to the global pandemic, which caused disruptions in students' lives and learning contexts. This disruption, too, likely influenced the response rate (50%). Yet this disappointment did not impact students' perception of what they gained: Survey results indicated students' positive perception of their experience and their own increase in language abilities and content knowledge. With only one exception, students either agreed (A) or strongly agreed (SA) with *all* statements, indicating their sense of achievement of LOs. Five items stood out with the highest ratings (see Table 9.3). As these items span LOs in both content and language, they indicate the breadth of learning students felt they achieved, from gaining the ability to apply business concepts to increasing their sustainability-related vocabulary.

While the survey served to gauge students' general perception of the extent to which they achieved the LOs, their reflective comments provided a richer understanding of their experience with the unit. An analysis revealed several larger patterns. Although they never used the term 'cross-disciplinary,' students commented positively about cross-disciplinary aspects of the unit, or, as one student worded it, "the interplay between the different disciplines." Indeed, the very wording of their comments often revealed that they had

Table 9.3 Students' self-assessment of learning outcomes. Items with highest agreement scores

After working with the sustainability unit, I am now able to...	*SWA*	*A*	*SA*
1 Present an example of my own ecologically conscious actions.	0	0	5
2 Compare/contrast sustainability perspectives and practices in Germany and the US.	0	0	5
3 Use data to analyze the current state of renewable energy in Germany.	0	1	4
4 Apply Hofstede's 'cultural dimensions' as a tool for analyzing the receptiveness to renewable energy in Germany and the approach of citizen energy cooperatives.	0	1	4
Working with the sustainability unit, I ...			
5 Expanded my vocabulary for understanding and discussing sustainability in German.	0	1	4

absorbed the notion of interweaving of the disciplines that we had intended when undertaking this curricular project. When students explicitly noted business-related content (or skills) that they felt they had mastered, their comment often situated this learning within the context of using (or expanding) their German language or cultural knowledge, such as "learning about the German methods of sustainable business in German." A mix of challenge and enjoyment went hand-in-hand with the mix of disciplines, as for several students who called cross-disciplinary aspects of the unit, such as "learning German terminology pertaining to sustainable business," both "most challenging" and "most enjoyable." Another student similarly labeled as "most engaging" yet also "most challenging" the experience of working with a business case in German, "because case studies in your native language can be challenging enough, let alone in your second language." Gauging by students' input, cross-disciplinarity so impacted their learning experience that language learning was not a separate element dictating course content; rather, it was integral to content-learning, as aimed for in CBI (Brinton et al., 2003).

A second trend in students' reflections was an emphasis on the high value of learning business content and tools during the sustainability unit. Two students saw learning how to discuss graphs and data as most valuable "specifically when learning a new language," because this "could be applied to a multitude of scenarios particularly in the business setting." Several identified Hofstede's cultural dimensions as "the most valuable aspect of the unit," since they could "understand the motivations of Germans and German businesses on a much deeper level." Others identified as "most enjoyable" and "fascinating" their learning to use Hofstede's framework to analyze the sustainability practices of the other culture "from a business perspective." Such reflections highlight the enormous potential of courses that link the study of language to other fields and content areas, e.g., business and sustainability, making the language skills transferable to contexts beyond the language classroom and into the 21st-century world with its demand for flexible competencies (King de Ramirez & Lafford, 2018; Long, 2017). The experience offered to students in this course may then help answer the call from the MLA report to ensure that language learning is relevant and connected to the new realities of a globalized economy, while simultaneously fostering the integration of liberal learning and business education as proposed by the Carnegie Report.

Finally, students' reflections on their experience revealed the unit's impact on their thinking about sustainability and their own role in it. Students credited the initial video and the business case with opening their eyes to German citizen activism. Through their 'greater awareness' of what German citizens are doing, they could step outside of their current cultural context (the US) and reflect on Germany's success in the energy transition, due in large part to citizen engagement, and see it as a challenge to be met rather than an unreachable bar (as expressed in one comment: "Citizen activism is growing here in the US but we have a long way to go, compared to Germany."). Seeing the impact of German citizen involvement in the context of sustainability changed students' thinking, which may inspire or motivate students to action. Developing sustainability literacy has the potential to promote in students the perspective of themselves as global citizens, which in turn can propel them to proactively take a position and ultimately act as stewards of the environment or even agents of change, as called for by scholars like Melin (2019) and Simpson and Mueller (2019).

Addressing Challenges in Implementation

There is great satisfaction in designing and delivering a course with the potential to impact the global concern of developing a sustainable future for our shared planet, particularly in the context of a business language course, where there are arguably heightened expectations for the practical and immediate applicability of students' newly gained competencies. Still, implementing this cross-disciplinary content-based approach brings challenges. In terms of content knowledge, business language instructors are already confronted with having to become knowledgeable about business content; adding sustainability content introduces yet another hurdle, demanding additional work that is rarely encouraged or rewarded by language programs and their institutions. Practitioners must therefore seek out opportunities to gain content knowledge conveniently and efficiently, through workshops and webinars within as narrowly focused an area as possible. They should look for a particularly obvious and robust nexus between sustainability and the culture(s) they teach, one that will provide an opportunity to analyze deeply held cultural values.

Working with colleagues from another discipline can complement gaps in content knowledge and free up the language instructor to focus on the effective integration of advanced language learning. The authors are fortunate to have a long-term ongoing partnership but realize this is not representative of what others might experience in the post-secondary context. More realistic would be to invite a colleague from the targeted discipline to give a brief presentation in class on a particular area of interest within one of the course themes. If this is must be done in English, a follow-up reading and discussion can help students draw a direct line from that presentation to their language and cultural learning. In our experience, finding a willing partner in the realm of sustainability-related teaching can be surprisingly easy, as they already realize the necessity of cross-disciplinary collaboration.

Even once a regular collaborative partnership is established, there are logistical hurdles to bringing the other faculty into your classroom with timing and location. Fortunately, pandemic-era teaching has demonstrated that online class visits are a viable option, demanding less time from a guest speaker and the option to record presentations. If the colleague can engage with students in the target language, this brings obvious advantages (additional language input, for example) but also potential roadblocks, depending on the colleague's proficiency level and tolerance of students' imperfections. The larger the collaborative commitment, the more hurdles cross-disciplinary partners will face, including department- and institution-level barriers to cross-disciplinary collaboration, which often make it difficult to cross-list courses, determine creative approaches to course loads, let alone recognize or reward the extra effort. Nevertheless, it is worth the initial effort to find a colleague with whom there is a synergy and who perceives the mutual benefit of working together.

Conclusion

In examining students' reflections, our priority was to gauge their response to both the topic and the approach, so that we can improve the unit for future iterations of the course. We would want to build on students' verbal acknowledgment that this coursework has changed their thinking, with the goal of transforming their words into concrete actions. We would encourage them to lean into the challenge of getting involved locally in sustainability actions through an assignment in which, for example, student teams explore their university's policies and procedures and propose an event or activity to improve these. To continue with a balanced CBI approach, such a task could integrate advanced language forms for convincing an audience of the value of sustainability work or recruiting others to get involved. Another variation of this project-based approach could be asking students to write a personal statement about their sustainability goals for the future, with commitment to undertaking specific action.

These changes would open up further avenues to evaluate student learning, e.g., through direct assessment of students' final writing projects.

The present project suggests the value of continuing curricular work in the context of a cross-disciplinary Business German course where the two disciplines converge on the timely—but also truly timeless—content focus of sustainability. Indeed, the approach we have presented here, purposefully linking Business Language Studies and sustainability, may now seem novel, yet we surmise that in the near future sustainability will be integral to all disciplines. Preparing our students to thrive in that cross-disciplinary future, within and across the languages and cultures that span our shared planet, is one of the most impactful contributions language educators can make.

References

American Council on the Teaching of Foreign Languages (ACTFL). (2012). ACTFL proficiency guidelines 2012. https://www.actfl.org/sites/default/files/guidelines/ACTFLProficiencyGuidelines2012.pdf

Becker, A., DeMaris, S. G., & Moller-Tank, B. (2013). Greening the German classroom: Starting points for a cultural lesson. *Die Unterrichtspraxis, 46*(2), 149–162.

Berg, B. M. (2013). Perspectives on the German *Energiewende:* Culture and ecology in German instruction. *Die Unterrichtspraxis, 46*(2), 215–229.

Bettencourt, M. (2011). Languages across the Curriculum: A response to internationalization in foreign language education. *Multicultural Education, 19*(1), 55–58.

Brinton, D. M., & Snow, M. A. (2017). The evolving architecture of content-based instruction. In M. A. Snow & D. M. Brinton (Eds.), *The content-based classroom: New perspectives on integrating language and content* (2nd ed., pp. 2–20). The University of Michigan Press.

Brinton, D. M., Snow, M. A., & Wesche, M. (2003). *Content-based second language instruction.* The University of Michigan Press.

Bundesministerium für Wirtschaft und Energie (BMWi). (n.d.). *Nachhaltigkeit in der Wirtschaft.* https://www.bmwi.de/Redaktion/DE/Textsammlungen/Wirtschaft/leitprinzip-nachhaltigkeit.html

Byrnes, H. (2005). Content-based foreign language instruction. In C. Sanz (Ed.), *Mind and context in adult second language acquisition* (pp. 282–302). Georgetown University Press.

Byrnes, H. (2013). German for specific purposes. In C. A. Chapelle (Ed.), *The encyclopedia of applied linguistics* (pp. 1–8). Blackwell Publishing Ltd.

Climate and Energy. (n.d.). *Facts about Germany.* FAZIT Communication GmbH. https://www.tatsachen-ueber-deutschland.de/en/climate-and-energy

Colby, A., Ehrlich, T., Sullivan, W., & Dolle, J.R. (2011). *Rethinking undergraduate business education. Liberal learning for the profession.* Jossey-Bass.

de la Fuente, M. J. (2019). Stepping out of the language box: College Spanish and sustainability. C. A. Melin (Ed.), *Foreign language teaching and the environment: Theory, curricula, institutional structures* (pp. 130–145). MLA.

Doyle, M. S. (2012). Business language studies in the United States: On nomenclature, context, theory and method. *Modern Language Journal, 96*(1), 105–121.

Ducate, L. (2016). Casting a university-wide net: Teaching sustainability in Germany. *Die Unterrichtspraxis, 49*(1), 69–79.

Forman, J., & Rymer. J. (1999). The genre system of the Harvard case method. *Journal of Business and Technical Communication, 13*(4), 373–400.

Fridays for Future. (2020, November 19). *Strike statistics.* https://fridaysforfuture.org/what-we-do/strike-statistics/

Gonglewski, M., & Helm, A. (2010). An examination of business case methodology: Pedagogical synergies from two disciplines. *Global Business Languages, 15,* 17–31.

Gonglewski, M., & Helm, A. (2013). LaissezFair: A case for greening the business German curriculum. *Unterrichtspraxis, 46*(2), 200–214.

Gonglewski, M., & Helm, A. (2014). Sustainability pedagogies for the business language classroom. *Global Business Languages, 19*, 3–21.

Gonglewski, M., & Helm, A. (2017). Crossing disciplinary lines to engage students in cross-cultural learning during short-term study abroad. *Global Advances in Business and Communications, 6*(1), 1–16.

Grosse, C. U., & Voght, G. M. (2012). The continuing evolution of languages for specific purposes. *The Modern Language Journal, 96*(s1), 190–202.

Hofstede, G. (2011). Dimensionalizing cultures: The Hofstede model in context. *Online Readings in Psychology and Culture, 2*(1), 3–26.

King de Ramirez, C., & Lafford, B. (2018). *Transferable skills for the 21st century. Preparing students for the workplace through world languages for specific purposes.* Sabio Books.

Lafford, B. A. (2012). Languages for specific purposes in the United States in a global context: Commentary on Grosse and Voght (1991) revisited. *The Modern Language Journal, 96*(s1), 1–27.

Long, M. K. (2017). Introduction. LSP Studies and the creation of translingual and transcultural competence. In M. K. Long (Ed.), *Language for specific purposes. Trends in curriculum development* (pp. 1–11). Georgetown University Press.

Long, M. K., & Uscinski, I. (2012). Evolution of languages for specific purposes programs in the United States: 1990–2011. *The Modern Language Journal, 96*(s1), 173–189.

Lyster, R. (2017). Language-focused instruction in content-based classrooms. In M. A. Snow & D. M. Brinton (Eds.), *The content-based classroom: New perspectives on integrating language and content* (2nd ed., pp. 109–123). The University of Michigan Press.

Melin, C. A. (2019). Introduction: Environmental thinking through language. In C. A. Melin (Ed.), *Foreign language teaching and the environment: Theory, curricula, institutional structures* (pp. 1–13). MLA.

Modern Language Association (MLA). (2007). Foreign languages and higher education: New structures for a changed world. https://www.mla.org/Resources/Research/Surveys-Reports-and-Other-Documents/Teaching-Enrollments-and-Programs/Foreign-Languages-and-Higher-Education-New-Structures-for-a-Changed-World

Ryshina-Pankova, M. (2013). Understanding "Green Germany" through images and film: A critical literary approach. *Unterrichtspraxis, 46*(2), 163–184.

Sacco, S. J., Jones, A. M., & Sacco, R. L. (2014). Incorporating global sustainability in the business language curriculum. *Global Business Languages, 19*, 23–35.

Sanchez-Lopez, L., Long, M. K., & Lafford, B. A. (2017). New directions in LSP research in US higher education. In M. K. Long (Ed.), *Language for specific purposes. Trends in curriculum development* (pp. 13–34). Georgetown University Press.

Sato, S., Hasegawa, A., & Kumagai, Y. (2017). Content-based instruction (CBI) for the social future: A recommendation for critical content-based language instruction (CCBI). *L2 Journal, 9*(3), 50–69.

Simpson, P. A., & Mueller, M. J. (2019). German is the new green? Language, environmentalism, and cultural competence. In C. A. Melin (Ed.), *Foreign language teaching and the environment: Theory, curricula, institutional structures* (pp. 17–38). MLA.

Takami, T. (2008). Employing international business cases for business language instruction. *Global Business Languages, 13*, 63–79.

United Nations. (n.d.). *Sustainable development goals.* https://www.un.org/sustain abledevelopment/sustainable-development-goals/

Weigert, A. (2004). "What's business got to do with it?" The unexplored potential of business language courses for advanced foreign language learning. In H. Byrnes & H. H. Maxim (Eds.), *Advanced foreign language learning: A challenge to college programs.* AAUSC Issues in Language Program Direction (pp. 131–150). Thomson Heinle.

Wieg, A. (2013). *Please, in my backyard: How renewable energy cooperatives advanced citizen involvement in the German energy transition.* Heinrich Böll Foundation. https://us.boell.org/en/2013/12/19/please-my-backyard-how-renewable-energy-cooperatives-advanced-citizen-involvement-german

10 Content-Based Instruction in a Spanish Language Classroom

Climate, Identity, and Historical Patterns of Latin American Migration to the United States

Silvia M. Peart, Bradford S. Barrett and Sharika D. Crawford

Introduction

In 2016, the Douglas Fir Group (DFG, 2016) called for a cross-disciplinary approach within the field of second language acquisition (SLA) to address and respond to new and complex societal and political problems. More recently, Wagner et al. (2019), with the support of the American Council on the Teaching of Foreign Languages (ACTFL), published a groundbreaking book where they advocate for the teaching of world languages that makes connections and collaborates with other disciplines in order to prepare students to address real-world challenges. These calls for multidisciplinarity in foreign language curricula are not new: The well-known 2007 Modern Language Association (MLA) report from the Ad Hoc Committee on Foreign Languages had called for greater collaboration among colleagues of all disciplines as one essential feature of new foreign language curricula. Since then, many language professionals in the field continue to champion interdisciplinary approaches across language courses (Lafford et al., 2014).

In the 2019–2020 academic year, responding to this call for a cross-disciplinary approach to world language teaching, the authors embarked on a collaboration that connected their respective disciplinary fields—applied linguistics, meteorology, and history. Their pedagogical goal was to develop a sustainable education instructional module that focused on the links between climatic shock events, historical patterns of migration, and identity. This module was intended for a university-level Spanish course. An emerging body of work was incorporated into the module that now indicates a link between migration and high-impact climate shock events (Feng et al., 2010; Hunter et al., 2015; Nawrotzki & DeWaard, 2016). It was essential to teach migration from a multidisciplinary perspective because until recently, most of the scholarship on migration had come from the humanities and social sciences (Durand et al., 2016; Rumbaut & Portes, 2001; Young, 2015). However,

DOI: 10.4324/9781003080183-14

human migration is now beginning to be seen as a consequence of a failure to attain sustainable agriculture practices, political environments, or urban development (Eakin et al., 2014; Rodriguez et al., 2018; Yang et al., 2020). Communities without those sustainable practices are now being seen as vulnerable to climate change and shock events, and migration is a consequence of that lack of sustainability (Wilkinson et al., 2016).

Given the urgency surrounding these issues, the authors thought this was a very relevant topic to be the focus of a multidisciplinary instructional module that used a content-based model of instruction (CBI) in Spanish. CBI helps learners acquire knowledge and expertise in new disciplines and develop greater cultural awareness, all through the target language. The instructional module introduced students to migration as a phenomenon with multiple sustainability dimensions. Students in this course are increasingly the children of such migrant families, yet the existing language curriculum has not adequately integrated their family stories and lived experiences. In centering the learning module around migration issues, the authors expected to validate and valorize the experience of heritage language learners (HLLs) while instilling empathy and cultural sensitivity to second language learners (SLLs). Using CBI, both HLLs and SLLs were exposed to language, topics, and concepts they may have never considered and thus learned about the content in the target language while acquiring new perspectives.

Content-Based Instruction (CBI) and Cross-Disciplinary Language Learning

CBI is well-grounded in Second Language Acquisition (SLA) research, such as the work of Krashen (1982, 1985), Swain (1985, 1995), and Cummins (1989). First, Krashen's (1982, 1985) argument that a second language (L2) is best acquired through extensive exposure to comprehensible L2 input supported the use of CBI, which was associated initially with immersion programs developed in Canada during the 1960s. In this context, Swain conducted extensive research, formulating her *output hypothesis* (Swain, 1985, 1993, 1995) in order to point out the limitations of these programs. As a result, she called for a more explicit focus on learners' production (as opposed to just comprehension) of language to support content-learning activities in the classroom. Another important support for CBI comes from Vygotsky (1986) and his sociocultural historical theory, particularly the concepts of negotiation in the Zone of Proximal Development and student appropriation of learning tasks. Grabe and Stoller (1997) underscored these contributions as relevant to understanding how learning occurs in L2 content-based classes. They argue that learners achieve a more significant cognitive and linguistic development through interaction with their peers and teachers than they would alone. The last theory construct supporting CBI is Cummins' (1989) notion of Cognitive Academic Language Proficiency (CALP). Cummins argues that many L2 students learn basic skills to communicate in the target language, but these are not sufficient to succeed in academic learning contexts. Cummins' work underscores the need for language educators to develop an inquiry-driven CBI approach (Cammarata, 2016) to engage learners cognitively while building their academic language repertoire.[1]

There are various forms of and approaches to CBI (Sato et al., 2017), depending on what counts as 'content' in various instructional contexts. Accordingly, several distinct types of CBI models have been proposed, such as languages across the curriculum (LAC), languages for specific or special purposes (LSP), and theme-based approaches based on how language instruction is structured vis-à-vis content instruction. In the case presented here, it is the language professor who is team-teaching and collaborating in a multidisciplinary setting with colleagues from history and climate science. The content chosen for this instructional module is challenging, relevant to the students' lives, and sometimes difficult and sensitive to discuss. Working with this complex and multidimensional content is critical for the language classroom. By engaging with it, students explain and share their thoughts and opinions with others in the target language and learn appropriate ways to share their newly acquired knowledge, thus making comparisons and interpretations so they can learn from each other.

In the subsequent sections, we offer a description of the instructional module: The higher education context in which it was taught, including a description of the students, and the instructional components and classroom learning activities or tasks. Analysis and discussion of students' reflections after two of the class sessions will follow.

The CBI Learning Module: Migration— Environmental and Human Sustainability

The module had four main learning goals: (1) Raising students' awareness about environmental and human sustainability by learning how climatic shock events are one of several reasons for Latin American migration into the US. In particular, it focused on Mexico and Central America as case studies, as those regions are connected to the Latino populations in the US; (2) enriching students' understanding of the relationship between climatic shock events and migration through small group and whole-class discussions; (3) promote students' critical thinking; and (4) promote language development. Students produced written reflections, research projects, oral presentations, and final exam responses, which we used to assess learning outcomes.

Context

Although the United States Naval Academy (USNA[2]) does not offer a major in Spanish; midshipmen in any major may pursue a minor in the language. The module was situated in an advanced Spanish elective course titled 'Latino/ as in the United States.' The instructor uses CBI throughout the course and

Table 10.1 Student participants in the CBI module

Majors of study	STEM field majors: Computer science, mathematics, mechanical engineering, physics, and quantitative economics
	Humanities majors: Arabic, English, political science
	One undeclared major
Age	Younger than 25 years old
Gender	Eight females; three males
Heritage language	Seven SLLs
	Four HLLs

teaches it entirely in Spanish to develop students' linguistic proficiency while advancing their subject knowledge. In particular, the course explores the historical forces that push and pull people from Latin America to the US. It also pays close attention to how a group known invariably as Spanish, Hispanics, Latinos, and Latinxs have adapted and resisted their positionality within the US. Immigration is a topic thoroughly covered throughout the semester.

Enrolled students were either HLLs or SLLs who were pursuing a minor in Spanish and desired to strengthen their linguistic proficiencies (see Table 10.1). Offering a learning module in this course had several advantages. The instructors in charge of the other disciplinary content (history and meteorology) had the required Spanish language skills, at the superior and advanced-high levels (based on the ACTFL scale), to support the course instructor in delivering the content of the module. In doing so, these non-language instructors served as models to other second language learners in the course, as they used the language in a professional setting. For these reasons, this course was an ideal setting in which to insert this curricular innovation.

Instructional Activities

The learning module comprised six one-hour class periods (see Figure 10.1). The first class period consisted of an hour-long introduction to the module, followed by 3 hour-long class periods. Each author led content and instructional activities, including succinct context presentations, video clips, and close textual reading of oral history sources. Each of these three class meetings ended with students writing reflection pieces on the content of the day. The fifth and sixth days of the module were devoted to 15–20-minute oral presentations by students on research projects they selected from a topic list given to them. The six classes were conducted in Spanish. Finally, two questions on the course final exam addressed content learned in the module.

Class 1: Introduction to the Interdisciplinary Module

The instructors presented the module and the learning objectives to the students. Since the module focused on migration and used it as the common

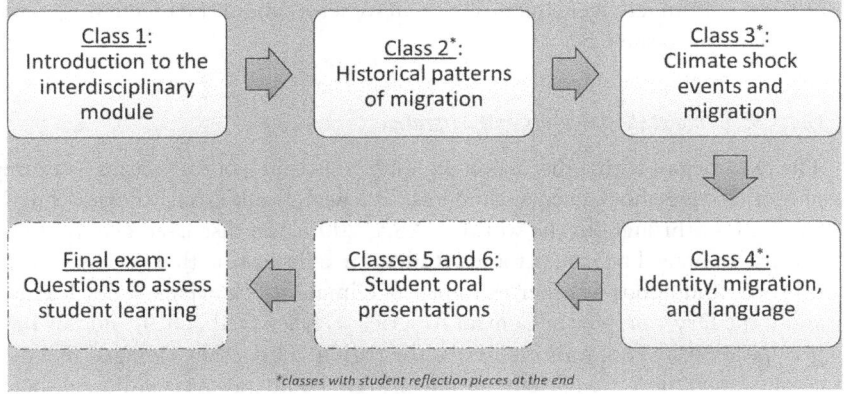

Figure 10.1 Flow chart summarizing the learning module

thread across the three fields of study, students could examine it from multiple perspectives and in the Spanish language. The authors presented a schedule of class activities, went over guidelines for student and instructor participation, and discussed a rubric to grade the oral presentation of a research proposal that would serve as the final product. The idea behind the module was to facilitate students' work across disciplines and scaffold students' understanding of content knowledge of these disciplines while also facilitating cultural and linguistic development.

Class 2: Historical Patterns of Migration

The goal was to frame the discussion about migration within a more extensive historical process. Drawing from the discipline of history, the instructor sought to reactivate previous learning and give students ample room to practice critical historical thinking skills. She asked students to consider the following overarching questions: Why did Mexicans move to the US, and how did US policies affect the flow of migration? To answer these questions, students engaged with original Spanish language textual sources related to Mexican migration to the US. Each source gave an authentic voice to a person's experience with migration. One source recounted a family's history in Nevada before and after the territory became a part of the US in 1848. Another source included excerpted oral histories of Mexican laborers who participated in the Bracero Program, an agreement between Mexico and the US governments to contract Mexican laborers to the US between 1944 and 1964. Finally, students analyzed a *retablo* (or a religious altarpiece image) that depicts a family's journey from Mexico to the US. Each source presented students with different reasons for the Mexican presence in the US. The class concluded with students preparing a written reflection on the following questions: What have been the patterns of Mexican migration to the US? What has changed over time? What can we learn about the recent migration of Mexicans to the US?

Class 3: Climate Shock Events and Migration

The class began with a discussion on hurricanes and tropical storms as examples of climate shock events. Students reviewed graphics of historical hurricane tracks throughout the world (NASA, 2006) and discussed geographical features of note. The instructor led students in a discussion that reviewed heat, drought, and floods as other examples of climate shock events. They learned about the Dry Corridor in Central America (Gotlieb et al., 2019) and saw how precipitation was not uniform across the region. Then, they engaged in a discussion about how a single climate shock event could have very different societal impacts, thus leading to migration, depending on the level of sustainability of socio-economic and political factors. The module returned to the idea of sustainability, adaptation, and migration: Unsustainable human-driven land surface changes (including urbanization and deforestation) have altered the ability of the natural ecosystem to absorb climate shock events, thus exacerbating the risks from those events and making migration a mitigation strategy. Students prepared written reflections on the questions: What are the natural hazards that Latin America faces? Are there ways to diminish vulnerability? Explain the relationship between climate change and migration in Latin America.

Class 4: Identity, Language, and Migration

In this class, students explored the relationship between identity, language, and migration. The instructor began by focusing on the relationship between identity and the concepts of space and place through a series of questions. Then she explained the concept of place attachment (Manzo & Devine-Wright, 2020), focusing on how a geographical space is also a social system where families and communities create networks of economic, social, and cultural relevance. Sometimes, migrants cannot replace or rebuild these networks in their place of arrival. The class also discussed what happens when people are forced to migrate, the trauma that this entails for the whole community and the individuals involved (Simms, 2008), and how this contributes to the dissolution of a community (Fullilove & Wallace, 2011). Finally, the instructor presented the 2019 documentary *Why Are Indigenous People Dying at The Border?*, which raised questions about how indigenous Guatemalan migrants reconstruct their identity through language. The class finished with a written reflection focused on the questions: How do migration and identity relate to each other? What problems do migrants face in the US in terms of identity and language?

The CBI Learning Module: Analysis of Students' Reflections

The primary materials we used in this study to assess the effectiveness of this interdisciplinary project were the written reflections completed by students after Classes 2, 3, and 4. These writings allowed students to display newly acquired knowledge, provide us with a window to their opinions and level of engagement with the topics and critical thinking, their writing skills, and especially their newly acquired, discipline-specific vocabulary. In addition to all of this, they allow us to obtain indirect evidence of the success of this pedagogical intervention.

Each reflection was assessed using a rubric (see Table 10.2) adapted from Crane and Sosulski (2020). In the Writing Ability category of the rubric the authors incorporated descriptors from the ACTFL's Writing Proficiency Guidelines (2012). Most expected writing learning outcomes correspond to the advanced range and a few to the superior range. These last ones require some scaffolding and structured support. In this level, students are expected, for example, to describe and narrate in extended paragraph-length discourse, express and support opinions, and write about a range of professional and complex topics. Reflection was also assessed using qualitative *thematic analysis* to identify salient themes discussed by learners each day. Research-derived codes (Crane & Sosulski, 2020) for each salient theme were identified by the lead author and corroborated by the others. The frequency of use of both the central theme and subthemes was calculated for each class reflection.

Reflection from Class 2

Table 10.3 shows a summary of how students performed on their reflection from Class 2. Most students (81.5%) met or exceeded expectations, and at least half of the students (54.5%) used paragraph-length constructions in the target language with precision and detail. They were able to develop arguments to support their opinions showing good control of a range of grammatical structures, and their writing exhibited a variety of cohesive devices. Thoughts were expressed clearly and supported by concrete examples. Students focused on migration-related policies, and 72% used discipline-specific vocabulary. For example, several wrote about the Bracero Program, Immigration Act of 1924 (referred to as '*Ley de cuotas*'), or the importance of having legal status (referred to as '*papeles*') (see Figure 10.2 for a summary of words used in the reflection).

About half of the students (45.5%) composed simple summaries linking sentences into short paragraphs. At times, students wrote sentence-level constructions using bullet points. In this case, their writing incorporated some organizational features using a limited number of cohesive devices, sometimes resorting to redundancy and repetition.

A thematic analysis of this reflection revealed that 72.7% of students explored the historical patterns of migration from Mexico to the US, focusing

Table 10.2 Rubric to assess students' written reflections

	Surpasses expectations	*Meets expectations*	*Below expectations*
Detail	• addresses prompt fully and with rich detail	• addresses prompt adequately with some detail; some points may require elaboration	• addresses prompt minimally with little or no detail
Specificity	• provides specific examples to illustrate and/or support ideas discussed	• provides some examples to illustrate and/or support ideas, but may be unspecified	• examples provided are vague and need specification
Engagement	• responds to class materials thoughtfully and critically	• responds to class material with some thought, but may lack critical awareness	• draws on class material superficially or minimally
Writing ability	• writes with significant precision and detail • uses paragraph length construction • can develop arguments and construct hypotheses with good control of a range of grammatical structures • exhibits a variety of cohesive devices • expresses thoughts clearly, supported by some elaboration	• incorporates some organizational features of the L2 and the L1; may at times resemble oral discourse • can compose simple summaries linking sentences into texts of paragraph length and structure • uses a limited number of cohesive devices, sometimes resorting to redundancy and awkward repetition • writes clearly most of the time; some mechanics may distract the reader	• uses sentence level or bullet points • does not use cohesive devices • writing style closely resembles oral discourse • shows little evidence of deliberate organization • writes clearly only some of the time
Lexical ability	• uses specialized vocabulary provided by the teaching module	• uses some specialized vocabulary provided by the teaching module	• uses little specialized vocabulary provided by the teaching module

on shifting US policies toward Mexican migrants. In contrast, students could have addressed the changing conditions surrounding Mexicans in the US. For example, students learned about Mexican families in Nevada who had to adjust when their territory transferred from Mexico to the US after the Mexican-American War of 1847. Alternatively, students could have discussed the effect of the Second World War and the US demand for labor on the recruitment of Mexican migrant workers. Additionally, the authors identified three subthemes that students analyzed and critically evaluated. First, 81.8% of students examined different US policies that may or may not have

Figure 10.2 Word cloud of student reflections after Class 2 on the history of migration

Table 10.3 Summary of student performance on reflection from Class 2

	Surpassed expectations	Met expectations	Below expectations
Overall	27%	54.5%	18%
Level of detail	45.4%	36.3%	18%
Specificity	45.4%	36.3%	18%
Engagement with class material	54.5%	27.2%	18.3%
Writing conventions	54.5%	45.5%	
Vocabulary	72.7%	27.3%	

provided a legal path for Mexicans to migrate to the US. Second, 63.6% of students discussed how US policies and even Americans held aggressive/ negative views towards Mexican immigrants. Finally, 45.4% of students described and synthesized how the United States implemented policies to welcome Mexicans when convenient. However, policies increasingly perceived Mexican migrants as problematic and have curtailed legal pathways for migrants to enter the US.

Reflection from Class 3

Table 10.4 shows a summary of how students performed on their reflection from Class 3. Most of the students surpassed (45.4%) or met (45.4%)

Figure 10.3 Word cloud of student reflections after Class 3 on migration and climate shock events

Table 10.4 Summary of student performance on reflection from Class 3

	Surpassed expectations	*Met expectations*	*Below expectations*
Overall	45.4%	45.4%	9.2%
Level of detail	45.4%	45.4%	9.2%
Specificity	45.4%	36.3%	18.3%
Engagement with class material	54.5%	36.3%	9.2%
Writing conventions	81.8%	17.3%	0.9%
Lexical ability	100%		

expectations. Most (81.8%) used paragraph-length constructions, providing thoughtful reflections with specific examples. They were able to develop arguments to support their opinions, showing good control of a range of grammatical structures (see Figure 10.3 for a summary of words used in the reflection). The word cloud shows that students used specialized vocabulary related to climatic shock events, geographic, and scientific terms, focusing on vulnerability and the different natural phenomena that can cause migration.

Overall, the reflection from Class 3 shows improvement in terms of writing conventions and paragraph length writing when compared with the reflection from Class 2. A thematic analysis of the reflection from Class 3 revealed that 100% of students evaluated the vulnerability of Latin America in terms of climatic shock events, exploring concrete examples in their reflections. This was the central theme in this reflection. In the first subtheme, 54.5% of students examined the importance of understanding and predicting natural disasters.

In the second subtheme, 45.4% of students connected climatic shock events to migration by giving concrete examples.

Reflection from Class 4

Table 10.5 reveals that all students either surpassed (72.7%) or met (27.2%) expectations. Overall, in their reflection from Class 4, more students exceeded or met the expectations than in the reflections from Classes 2 and 3. In terms of linguistic performance, more students (90.9%) used paragraph length constructions producing reflections that critically engaged with the material presented in class. They were able to develop arguments to support their opinions, showing good control of a wide range of grammatical structures. Their writing exhibited various cohesive devices, and their thoughts were expressed clearly and supported by some elaboration and concrete examples.

For this reflection, students used discipline-specific vocabulary, even connecting migration with climatic shock events (see Figure 10.4 for a summary of words used in the reflection). The word cloud shows that students explored topics related to identity, language, culture, and land *(identidad, lengua, cultura, tierra)*, all of which were discussed in Class 4.

Overall, students performed better in the reflection after Class 4, showing improvement in language mechanics and paragraph-length writing

Table 10.5 Summary of student performance on reflection from Class 4

	Surpassed expectations	*Met expectations*	*Below expectations*
Overall	72.7%	27.2%	
Level of detail	54.5%	45.5%	
Specificity	54.5%	45.5%	
Engagement with class material	63.6%	36.4%	
Writing conventions	90.9%	9.1%	
Vocabulary	72.7%	27.2%	

Figure 10.4 Word cloud of student reflections after Class 4 on identity and migration

compared with the reflections after Classes 2 and 3. A thematic examination of the reflection after Class 4 showed that the central theme was the critical analysis of the relationship between place, language, and identity, with concrete examples. All students (100%) explored this topic on their reflection. Additionally, the authors identified three subthemes. First, 81.8% of students demonstrated an understanding of the ways linguistic barriers impede immigrants from fully participating in their new community. Second, 54.5% explored and emphasized the deep sense of loss that migration can cause. Finally, 45.4% expressed an understanding that indigenous groups have unique challenges on the border due to linguistic and cultural barriers, unlike their Spanish-speaking counterparts. Finally, students wrote more for the reflection after Class 4; that is, they produced more language, 828 words in the reflection after Class 2 versus 1,062 in the reflection after Class 4.

Discussion

In this module, taught in a content-based language class where students use Spanish in authentic and meaningful ways to develop proficiency, the topic of migration—treated from a sustainability perspective—brought together different disciplinary perspectives, which allowed students to engage with the subject in various ways. Despite the multifaceted nature of the topic, the existing scholarship on Latin American migration to the US privileges economic and political reasons. It considers migration to be a response to unsustainable elements. In doing so, other factors such as climate and climatic shock events are not yet fully incorporated (Feng et al., 2010; Nawrotzki & DeWaard, 2016). By developing this cross-disciplinary instructional module, the authors exposed students to a much-needed socio-environmental perspective of migration, so they could learn and reflect upon difficult yet realistic situations, including the formal and informal processes of migration, the risks of climatic shock events, and the ways migration affects social, cultural, and linguistic identities. The module prioritized students' use of the newly acquired content knowledge in Spanish to understand the topic of Latin American migration to the US and develop *critical thinking* skills. Let us briefly examine how the results of each CBI class contributed to the success of the cross-disciplinary module.

Class 2

The instructors' carefully scaffolded activities provided the necessary tools for students to produce complex language, including discipline-specific vocabulary. While only 45.5% of students produced language at the paragraph level, they could produce complex thoughts in simple language. For example, one student said, "laws change over time, and papers become more important." Students noted the growing importance of *papeles* (papers) as immigration became increasingly regulated through a legal process of written documentation. Although the language they used was simple, students established historical patterns, understanding the nuances of complex historical processes, and wrote about them in Spanish. On the other hand, 54.5% of students could write paragraph-length discourse about historical and cultural topics expressing and supporting their opinions, making connections with historical events not directly covered in Class 2 but previously presented in the course. For example, nearly two-thirds of the responses discussed the Bracero Program, as is evident in the word cloud (see Figure 10.2). The consistency of its inclusion in the reflections suggests that the students recognized it as a critical policy in regulating legal entry into the US.

Yet, the Bracero Program was not the only law referenced in the reflections. Students also noted the Immigration Act of 1924 (referred to as '*Ley de cuotas*'), which enforced a quota system for immigrants seeking entry from numerous countries. As several students noted, Mexico was excluded from the quota system, and migrants experienced fewer entry restrictions. Students' close attention to the shifting laws to enter the US for Mexican migrants allowed them to infer claims about those historical patterns. In analyzing the consequences of the Bracero Program's demise in 1964, one student argued that it became "much more difficult for Mexicans to migrate legally and more punishments for undocumented immigrants." Another student explained how "the Immigration Act of 1965 removed the quota system of 1924. But the migrants were not working under the official program, and they had fewer protections after 1965." As a result, students empathized with the plight of Mexican migrants while noting an essential historical thinking skill—continuity and change over time. One student wrote, "In general, the attitude about immigration had changed to be more aggressive." Another student shared a similar sentiment, "In general, immigration from Mexicans between the United States had been very negative." These examples demonstrate how students developed historical thinking skills while using the target language in sophisticated and meaningful ways and employing policy-specific vocabulary.

Class 3

Students' reflections highlighted that different climate shock events bring risk, and they were able to name several types of shock events, like hurricanes, droughts, and floods, as shown in the word cloud represented in Figure 10.3. All students were able to recall and accurately use discipline-specific vocabulary. Reflections showed more students writing using paragraph-length constructions when compared with Class 2 (81.8% vs. 54.5%). We hypothesize that this improvement in students' performance may result from more familiarity with this task. This was their second reflection at the end of the class, so perhaps students were able to anticipate this task, something they could not the first time around. Finally, it is worth noting that with careful scaffolding activities and providing the tools necessary to complete the tasks, students can perform at a higher level in specific tasks because they collect the necessary information in advance (Wagner et al., 2019).

Students mentioned some of the major climate-driven reasons for people to migrate. For example, two students tried to explore regional differences in climate risk and wondered whether one particular place was "better" than another: "Many try to live in a better place" and "the (climate) risks force people out of their country and to better places like Florida." These interpretations suggest that students believe that migration linked to climate shock events is toward a "better" place. Two other students defined prevention measures: "Several countries can prepare for the possibility of a climate (shock) event" and "one way to improve conditions after a natural 'shock' is to predict a disaster." This discussion shows the ways students employed new vocabulary and content on the topic of climatic shock events and migration, implying that predictions would warn of a shock event, allowing communities to prepare and thus mitigate the possible impacts and the migration response. The reflections evidence students' realization that migration is a consequence of people searching for a more sustainable place in the face of climate risks.

Students could complete linguistically sophisticated written work, stretching their linguistic proficiency and conceptual knowledge about climatic shock events, connecting these ideas with the overarching topic of migration. Students evaluated geographical risks or drought around Latin America, focusing mainly on the Dry Corridor in their reflections. They recognized and expressed, in writing, different threats from the climate system and realized that these threats were unevenly distributed geographically in Latin America.

Class 4

Students, both HLLs and SLLs, wrote very personal reflections about identity and language, linking what they learned to their lives beyond the classroom, which suggests that class discussions allowed them to gain new perspectives and ideas from other learners. In terms of linguistic proficiency, the fact that more students used paragraph-length constructions to explore complex ideas than in Classes 2 and 3 (90.9%) could be attributed to (1) the motivation derived from working on a relevant topic that corresponded with their personal interests, with the opportunity to write about personal experiences; and (2) the type of questions asked, which may have encouraged more opinion and interpretation on the part of the students. In their reflections, learners had to go beyond retrieving and analyzing newly acquired information; they also had to think reflectively, explain, and justify their thoughts. This process, while cognitively challenging, may have promoted engagement with the reflective writing process.

Reflections analyzed the particular challenges that indigenous people have faced, highlighting three important takeaways: They recognized an important distinction between immigrant and refugee, which does not always translate into legal benefits; they emphasized the sacred connection with the land and the long struggle that indigenous peoples have faced with keeping their lands; and they also underscored the importance of language, pointing out that Spanish and English were both colonial languages, although Spanish does not have the same value as English in the US.

Finally, it is worth noting that students identified and explored how simplistic stereotypes in the receiving nation complicate migrants' reconstruction of their new identity. One student commented, "Americans generalize all immigrants from Latin America and call them Latinos." Another student also underscored how misperceptions about immigrants affect that process,

> One example of the challenges immigrants from Mexico have to face is that they left places of violence and corruption and when they arrive in the U.S. they are called murderers and rapists. This makes it harder for them to accept this place as part of their identity.

These examples show that students understood and evaluated the issue from multiple perspectives, showing empathy for others and expressing complex ideas using the target language in a sophisticated manner.

Conclusion

This chapter analyzed the impact of a multidisciplinary instructional module that incorporated sustainable development education through the topic of migration, climatic shock events, and cultural identity in an undergraduate-Spanish language course. The authors believe this module represents one effective model of how foreign language study in higher education should look like in the 21st century. In this instructional sequence, students

1   read, interpreted, and discussed original accounts of Mexican migrants to the US
2   collaborated with classmates to discuss, and later present, their ideas to the class on how place shapes their own sense of cultural identity
3   consistently integrated new Spanish language vocabulary such as geographic, scientific terms
4   integrated content related to historical patterns of and policies about migration
5   engaged in meaningful reflection and cross-disciplinary critical thinking
6   produced complex target language

This type of curricular work did not come without challenges. As with any multidisciplinary work, the authors found it challenging to explicitly link their content to the other units in the module. Without an opportunity to retrieve and reapply past content, the students could not work through an essential process of retrieval, synthesis, and integration of content. It would be necessary to extend the length of the module to accomplish these goals. A full-term course could also be developed that integrated all disciplinary components of the module under a sustainable education framework. Such a course would serve students well. This is linked to another challenge: The authors' institution does not offer a formal structure for team-taught classes (taught by multiple instructors across different departments). As a result, the module was situated in one of the author's disciplinary courses, and the other two authors had to design, deliver, and assess the module as an 'overload.' Although the module was delivered over two weeks, the curricular work entailed over 75 hours.[3] Thus, instructors looking to pursue similar innovation projects need access to supervisory and financial resources.

Perhaps the most rewarding aspect of the project for us, authors and instructors, was the use of much of our meetings to learn from each other's disciplinary perspectives, including the various methodological and theoretical approaches to the questions surrounding migration, climate, and identity. In doing so, a group of undergraduate students in a Spanish language elective course engaged with a relevant topic such as migration—in intellectually demanding ways—in the context of global environmental change.

Notes

1 For a look at successful CBI curricula, see Klahn (1997), Klee and Tedick (1997), Vines (1997), Cammarata (2010), and Byrnes (2019).
2 The United States Naval Academy (USNA) is one of three military service academies under the Department of Defense. USNA graduates baccalaureate students (known as midshipmen) who commit to serving in the US Navy or Marine Corps a minimum of five years as military officers in exchange for their no-cost education. The school mandates that 65% of all students major in a STEM field.
3 The authors were fortunate to receive an instructional innovation grant (N0001419WX01782) from the Office of Naval Research to spend part of their 2019 summer working collaboratively on these tasks.

References

ACTFL. (2012). ACTFL proficiency guidelines. https://www.actfl.org/sites/default/files/guidelines/ACTFLProficiencyGuidelines2012.pdf

Byrnes, H. (2019). Affirming the context of instructed SLA: The potential of curricular thinking. *Language Teaching Research, 23*(4), 514–532. doi: 10.1177/1362168818776666

Cammarata, L. (2010). Foreign language teachers' struggle to learn content-based instruction. *L2 Journal, 2*(1), 89–118. doi: 10.5070/L2219063

Cammarata, L. (Ed.). (2016). *Content-based foreign language teaching: Curriculum and pedagogy for developing advanced thinking and literacy skills.* Abingdon: Routledge.

Crane, C., & Sosulski. M. J. (2020). Staging transformative learning across collegiate language curricula: Student perceptions of structured reflection for language learning. *Foreign Language Annals, 53*(1), 69–95. doi.org/10.1111/flan.12437

Cummins, J. (1989). A theoretical framework for bilingual special education. *Exceptional Children, 56*(2), 111–119. https://doi.org/10.1177/001440298905600203

Douglas Fir Group. (2016). A transdisciplinary framework for SLA in a multilingual world. *The Modern Language Journal, 100*(S1), 19–47. doi: 10.1111/modl.12301

Durand, J., Massey, D. S., & Pren, K. A. (2016). Double disadvantage: Unauthorized Mexicans in the US labor market. *The Annals of the American Academy of Political Science and Social Science, 666*(1), 78–90. doi: 10.1177/0002716216643507

Eakin, H. C., Lemos, M. C., & Nelson, D. R. (2014). Differentiating capacities as a means to sustainable climate change adaptation. *Global Environmental Change, 27*, 1–8. doi: 10.1016/j.gloenvcha.2014.04.013

Feng, S., Krueger, A. B., & Oppenheimer, M. (2010). Linkages among climate change, crop yields, and Mexico-US cross-border migration. *Proceedings of the National Academy of Sciences, 107*(32), 14257–14262. doi: 10.1073/pnas.1002632107

Fullilove, M. T., & Wallace, R. (2011). Serial forced displacement in American cities, 1916–2010. *Journal of Urban Health, 88*(3), 381–389. doi: 10.1007/s11524-011-9585-2

Gotlieb, Y., Pérez-Briceño, P., Hidalgo, H., & Alfaro, E. (2019). The Central American Dry Corridor: A consensus statement and its background. *Revista Yu'am. Revista Mesoamericana de Biodiversidad y Cambio Climático, 3*(5), 42–51.

Grabe, W., & Stoller, F. L. (1997). Content-based instruction: Research foundations. In: M. A. Snow & D. M. Brinton (Eds.), *The content-based classroom: Perspectives on integrating language and content* (pp. 5–21). White Plains, NY: Longman.

Hunter, L. M., Luna, J. K., & Norton, R. M. (2015). The environmental dimensions of migration. *Annual Review of Sociology, 41*, 377–397. doi: 10.1146/annurev-soc-073014-112223

Klahn, N. (1997). Teaching for communicative and cultural competence: Spanish through contemporary Mexican topics. In S. B. Stryker, & B. L. Leaver (Eds.), *Content-based instruction in foreign language education: Models and methods* (pp. 203–221). Washington DC: Georgetown University Press.

Klee, C. A., & Tedick, D. J. (1997). The undergraduate foreign language immersion program in Spanish at the University of Minnesota. In S. B. Stryker & B. L. Leaver (Eds.), *Content-based instruction in foreign language education: Models and methods* (pp. 141–173). Washington DC: Georgetown University Press.

Krashen, S. (1982). *Principles and practice in second language acquisition*. Oxford and New York: Pergamon Press.

Krashen, S. (1985). *The input hypothesis: Issues and implications*. New York: Longman.

Lafford, B. A., Abbott, A., & Lear, D. (2014). Spanish in the professions and in the community in the US. *Journal of Spanish Language Teaching, 1*(2), 171–186. doi: 10.1080/23247797.2014.970361

Manzo, L., & Devine-Wright, P. (Eds.). (2020). *Place attachment: Advances in theory, methods, and applications*. Abingdon: Routledge.

MLA Ad Hoc Committee on Foreign Languages. (2007). Foreign languages and higher education: New structures for a changed world. *Profession, 2007*(1), 234–245.

NASA. (2006). Historical tropical cyclone tracks. https://earthobservatory.nasa.gov/images/7079/historic-tropical-cyclone-tracks

Nawrotzki, R. J., & DeWaard, J. (2016). Climate shocks and timing of migration from Mexico. *Population and Environment, 38*(1), 72–100. doi: 10.1007/s11111-016-0255-x

Rodriguez, R. S., Ürge-Vorsatz, D., & Barau, A. S. (2018). Sustainable development goals and climate change adaptation in cities. *Nature Climate Change, 8*(3), 181–183. doi: 10.1038/s41558-018-0098-9

Rumbaut, R. G., & Portes, A. (2001). *Ethnicities: Children of immigrants in America.* Oakland: University of California Press.

Sato, S., Hasegawa, A., Kumagai, Y., & Kamiyoshi, U. (2017). Content-based instruction (CBI) for the social future: A recommendation for critical content-based language instruction (CCBI). *L2 Journal, 9*(3), 50–69. doi: 10.5070/L29334164

Simms, E. M. (2008). Children's lived spaces in the inner city: Historical and political aspects of the psychology of place. *The Humanistic Psychologist, 36*(1), 72–89. doi: 10.1080/08873260701828888

Swain, M. (1985). Communicative competence: Some roles of comprehensible input and comprehensible output in its development. In S. M. Gass & C. G. Madden (Eds.), *Input in second language acquisition* (pp. 235–245). Rowley, MA: Newbury House.

Swain, M. (1993). The output hypothesis: Just speaking and writing aren't enough. *Canadian Modern Language Review, 50*(1), 158–164.

Swain, M. (1995). Three functions of output in second language learning. In G. Cook and B. Seidlhofer (Eds.), *Principles and practice in applied linguistics* (pp. 125–145). Oxford: Oxford University Press.

Vines, L. (1997). Content-based instruction in French for Journalism students at Ohio University. In S. B. Stryker & B. L. Leaver (Eds.), *Content-based instruction in foreign language education* (pp. 118–139). Washington, DC: Georgetown University Press.

Vygotsky, L. (1986). *Thought and language.* Cambridge, MA: The Massachusetts Institute of Technology Press.

Wagner, M., Cardetti, F., & Byram, M. (2019). *Teaching intercultural citizenship across the curriculum: The role of language education.* ACTFL (American Council on the Teaching of Foreign Languages). Alexandria, VA: National Education Association.

Wilkinson, E., Schipper, L., Simonet, C., & Kubik, Z. (2016). *Climate change, migration and the 2030 Agenda for Sustainable Development* (pp. 201–215). London: Overseas Development Institute.

Yang, Y., Liu, B., Wang, P., Chen, W. Q., & Smith, T. M. (2020). Toward sustainable climate change adaptation. *Journal of Industrial Ecology, 24*(2), 318–330. doi: 10.1111/jiec.12984

Young, J. (2015). *Mexican exodus: Emigrants, exiles, and refugees of the Cristero war.* Oxford: Oxford.

11 Sustainability across the Curriculum

A Multilingual and Intercultural Approach

Deborah Reisinger, Sandra Valnes Quammen, Yan Liu and Edgar Virgüez

What Is CLAC?

The Cultures and Languages Across the Curriculum framework espouses an innovative approach to integrating the study and use of foreign languages in departments across institutions, outside language departments. CLAC's philosophy is grounded in pedagogical movements that include various 'across the curriculum' efforts, as well as high-impact practices that encourage meaningful engagement and applied learning (CLAC Consortium, 2019). CLAC programs have existed in US institutions of higher education for nearly 40 years. With financial support from US Department of Education grants in the 1980s, these curricular initiatives—baptized as Foreign Languages Across the Curriculum (FLAC) programs—were designed to increase opportunities for students to use languages other than English in their courses of study. Later termed Languages Across the Curriculum (LAC), the purpose was to give students with intermediate levels of proficiency the opportunity to use their skills outside of conventional contexts. Today's programs generally fall under the nomenclature of CLAC, which reflects a renewed emphasis on culture (e.g. University of Utah 2013, Duke University 2014, University of Denver 2015). Once a small organization composed primarily of its founding members, the CLAC Consortium has doubled in size over the last 15 years, adding nine new institutional members in the past six years to include a total of 28 institutions. As Reisinger (2018, p. 18) notes,

> CLAC's growth dovetails with the MLA's oft-cited 2007 report that called for a shift in the practices of teaching languages in post-secondary institutions. In its recommendations for attracting students from outside traditional fields, for instance, the report cites CLAC directly, suggesting that 'a credit-bearing discussion module taught in the target language can be added with the support of programs such as foreign languages across the curriculum'.
>
> (Geisler et al., 2007, p. 239)

DOI: 10.4324/9781003080183-15

CLAC programs embrace the bold idea that languages other than English can and should be used throughout the curriculum instead of being confined to the 'foreign language classroom' or 'the language department.' As Zilmer writes (2018, p. 10),

> CLAC is committed to the concept that inclusion of other languages and cultures in content area studies provides access to knowledge, skills, and cognition that are not available by any other means. CLAC is content area learning that is accessed through the inextricable combination of language, culture, and cognition.

In this vein, CLAC practitioners and their programs draw on students' language abilities to add intercultural perspectives to a broad set of content areas (history, global health, environmental studies, etc.), whether these students are heritage speakers who grew up speaking the second language (L2), native speakers who are international students on our campuses, or language learners who began their language studies when they arrived at their institution. Rather than replacing traditional language learning, CLAC's integrative learning programs build on students' abilities by creating opportunities and motivation for continued use and development of linguistic and cultural proficiencies. This approach aligns with the global learning goals espoused by many institutions of higher education (Hovland, 2014), and in particular the authors' own, which recently articulated its strategic plan to include strengthening its "capacity to address global challenges for communities across the world" (Duke Strategic Plan).

CLAC imagines a campus where multiple languages are spoken across varied disciplines, bringing new, global research perspectives to projects often approached through materials produced uniquely in English. A sociology class on immigration that is taught in English, for instance, can become a transformative learning experience for students when paired with a CLAC section in Spanish; in this curricular environment, international students and heritage speakers work alongside language learners to interview Central Americans who have crossed borders and navigated US immigration policies. Likewise, a course in global health is enriched when paired with a CLAC tutorial in Mandarin that explores China's approaches to public health, reframing what might otherwise be deemed 'alternative medicine.' These efforts to value a diversity of voices and perspectives help to internationalize disciplines such as health, policy, and sustainability, creating opportunities for students and faculty to engage in perspective-shifting as they develop communicative competencies (Reisinger et al., 2015).

At the authors' institution, the CLAC program has fostered deep interdisciplinary connections between language studies and the schools of Global Health, Public Policy, and Environmental Sciences by pairing core courses within these schools with tutorials that work with complementary content in languages other than English. The program was founded in 2014 and has since offered 45 tutorials taught in nine different languages to over 460 students. While each CLAC tutorial is shaped by its core field, all classes are defined by the following four pillars:

- meaningful engagement with local and global populations
- exploration of culturally-specific solutions to real-world issues
- development of discourse competence for subject-specific use
- critical reflection about how language and culture impact worldview

CLAC tutorials are designed as half-credit courses, lending themselves to dynamic and flexible pedagogies that can respond to developing curricular needs or leverage nascent interdisciplinary collaborations. They have forged important community connections with local organizations, become full-credit courses, and spawned larger research and study away projects. In this way, CLAC serves as an incubator for innovative pedagogies that connect students with real-world issues in both local and global communities.

In this chapter, we focus on three CLAC tutorials housed within Duke University's Nicholas School of the Environment that were designed for high-intermediate and advanced-level learners and speakers of French, Mandarin Chinese, and Spanish. These tutorials were created in conjunction with a gateway course in the environmental studies major (ENVIRON 201). Each CLAC tutorial met weekly for 75 minutes and was structured around an in-depth engagement with case studies that link the essential topics in sustainability presented in ENVIRON 201 to the study of language and culture. The tutorials were overseen by the CLAC director and taught concurrently by a French language instructor, a Mandarin Chinese language instructor, and a Spanish native speaker content specialist (all four authors of this chapter). With this approach, we aim to foreground the essential role that linguistic and cultural competence play in sustainability education.

Developing Voices in the Environment

As comprehensive undergraduate programs in sustainability have become more widespread and popular in recent years, sustainability educators have elaborated a core set of competencies designed to serve as a framework for program design and educational outcomes (Clark, 2016; Wiek et al., 2011). These key competencies recognize the interdisciplinary nature of the field and provide reference points for articulating the broad goals of a discipline that aims to help students develop the knowledge, skills, and attitudes to become "future problem solvers, change agents, and transition managers" (Wiek et al., 2011, p. 204). Significant synergies exist amongst a number of these key competencies and the four pillars of the CLAC framework outlined above. Indeed, CLAC methodologies provide a venue in which to center the myriad voices involved in responsible, equitable change and transition toward more sustainable communities. For example, *systems-thinking* competence (defined as "an intimate understanding of the inner fabric and dynamics of complex social-ecological systems" (Wiek et al., 2011, p. 207)) and *interpersonal* competence (defined as "the capacity to understand, compare, and critically evaluate different positions, perspectives, and preferences" (Wiek et al., 2011, p. 211)) inform both sustainability education in the environmental sciences and the work and goals of CLAC tutorials that highlight the necessity of engaging respectfully and appropriately with stakeholders that represent diverse perspectives, priorities, and needs. Furthermore, Clark's (2016) concept of *collective action* competence, or "the capability of a group of people to direct their behavior toward a common goal based on a collective literacy... and a collective need" can help to underscore both the cultural specificity of sustainability actions undertaken by communities at a local level and the intersections of collective actions on a broader scale (p. 560).

Integrating Environmental Science and Policy (ENVIRON 201) is the gateway course for students pursuing a major in Environmental Sciences & Policy, and as such, it attracts students relatively early in their formal coursework towards the major. The course is designed to foster the development of key sustainability competencies as students are introduced to the complex interactions between natural and social systems in relation to the environment. ENVIRON 201 is constructed around four core modules, each representing a 'wicked problem' in environmental science. Wicked problems are environmental problems that are difficult or impossible to solve due to the number of stakeholders involved and their often-conflicting positions, and the absence of a single, definitive solution that applies in all cases and places (DeFries & Nagendra, 2017; Rittel, 1973). Students in ENVIRON 201 explore four such wicked problems—environmental justice, climate change and wildfires, hydropolitics, and biodiversity—that are at once global in scale and tightly linked to specific political, environmental, and social factors at a local level. Given the natural overlaps between the scope and framework of ENVIRON 201 and CLAC methodologies more broadly, in Fall 2020, the authors embarked on the pilot phase of pairing this gateway course with a cluster of CLAC tutorials offered in Mandarin Chinese, French, and Spanish entitled 'Voices in the Environment.'

In designing these CLAC tutorials to align with the modular structure of the ENVIRON 201 course, our group of faculty leveraged the interdisciplinary nature of our fields to develop a basic structure and instructional goals for these tutorials that were well-adapted to our language learners' linguistic and cultural proficiency, as well as their previous familiarity with sustainability-related topics. One author's expertise in CLAC theory and interdisciplinary course design provided the foundation of the tutorials' structures. Another author's disciplinary knowledge of environmental science ensured the relevance of the language- and culture-specific work with 'wicked problems.' Two authors' expertise in language pedagogy provided sound pedagogical grounding for effective student engagement with the course content. Finally, frequent input from the instructor of the ENVIRON 201 core course allowed the alignment of contents of the CLAC tutorials and the gateway course. The result of this collaborative process was an instructional model founded on three components: (1) Case studies as a means to engage with diverse voices and perspectives, (2) project-based inquiry, and (3) community engagement.

The three CLAC tutorials adopted a common set of learning outcomes that both described appropriate linguistic outcomes for intermediate-high level students and took inspiration from the Global Learning benchmarks detailed in the Association of American Colleges & Universities' (AAC&U) VALUE (Valid Assessment of Learning in Undergraduate Education) rubrics.[1] These included the ability to: (1) Identify environmental challenges in China, Latin America, the Caribbean, or Francophone regions; (2) discuss with relative ease, confidence, and complexity topics related to the themes of the course; (3) prepare and deliver short written and oral presentations on sustainability-related topics; (4) articulate an awareness of ways in which geography, history, language, and culture underpin the understandings of and reactions to sustainability-related issues in these regions; and (5) apply this awareness to create public-facing documents that are thematically-relevant and culturally-appropriate. Enrollment in the gateway environmental science course was encouraged but not required, and though the tutorials were scheduled with concurrent enrollment in mind, only a handful of students enrolled in both a CLAC tutorial and the ENVIRON 201 course. As the semester began, most students indicated having a keen personal interest in sustainability, and some had done previous sustainability-related coursework, though very few students were pursuing a major in environmental studies or a related field.

Case Studies

Given this student profile, we explored each of ENVIRON 201's four core modules via one to two case studies designed to both provide an introduction to fundamental facts and concepts related to each wicked problem, and to ground the specific linguistic and cultural factors at play.[2] Each module began with a guiding question or questions related to the topic at hand. Thematic materials produced in the target language were sourced from a variety of text types and stakeholder voices and included, for example, documentary videos and interviews, social media posts, newspaper articles, government reports, published case studies and policy briefs, and NGO websites. The faculty provided a series of comprehension and reflection questions to guide students' engagement with the materials in each thematic unit and to provide appropriate linguistic and lexical scaffolding. As they engaged with these materials, students collaborated asynchronously prior to our weekly sessions to create lists of fundamental concepts and terminology, adding to an evolving sustainability lexicon. During weekly synchronous discussions, students synthesized their knowledge by identifying stakeholders' positions and perspectives, analyzed the linguistic and cultural factors at work in each case, and even brainstormed potential solutions to environmental concerns.

Though each of the three CLAC tutorials explored the same four wicked problems, the specificity of our various thematic modules emphasized the importance of exploring these questions from local perspectives. For example, our collective exploration of a broad question such as 'How do history, language and culture inform environmental justice?' led to related but unique considerations in each of our linguistic contexts. In the French CLAC tutorial, for instance, students and instructor worked together to assemble a collection of documents related to the environmental and public health impacts of the use of the pesticide Kepone in banana cultivation in Guadeloupe and Martinique. Working with news segments, government reports, activists' social media presences, and press kit documents from Jessica Oublié's documentary graphic novel *Tropiques toxiques*, students explored how the region's colonial past continues to impact local populations and environmental policy, and discussed responses that would address the problem from both environmental and equity standpoints. Public health and the environment also intersected in the Mandarin Chinese CLAC tutorial's environmental justice module, which focused on cancer villages in rural China. These discussions centered the voices and positions of a variety of stakeholders, from villagers and local enterprises to NGOs, health experts, and the government, to trace the roots of the problem and brainstorm interventions and solutions. In the Spanish CLAC tutorial, students worked with news articles and government reports on asbestos use in Colombia to understand the reasons that prevented the government from banning the use of this carcinogenic mineral fiber. By exploring the disproportionate effect on automotive repair shop workers who are exposed to asbestos fiber and interacting with a local expert who had documented hundreds of exposures, students learned that a multidisciplinary approach to the problem is necessary to formulate efficient policies that protect workers' rights. Table 11.1 provides a brief summary of the linguistically- and culturally-specific themes that were explored in the three tutorials for each of ENVIRON 201's 'wicked problems.'

Project-Based Inquiry

Project-based language learning facilitates "academic discourse socialization, decision-making, critical thinking, and collaborative work skills while providing deep engagement with subject matter content through the use of language as medium" (Beckett et al., 2019, p. 8), and is therefore particularly well-suited to the CLAC model. In composing these modules, faculty modeled the type of reflective engagement with diverse perspectives and sources that students would need to demonstrate in each of the two projects that formed the basis of our assessment model. For their first project, completed at the midterm, pairs of students worked to identify a sustainability topic within one of the four core modules, and to prepare, in consultation with their instructor, a case study of their own related to this topic via a collection of texts and guiding questions. Students then presented their case studies to their classmates

Table 11.1 Thematic materials related to each of the four 'wicked problems'

ENVIRON 201 'wicked problems'	French tutorial themes	Chinese tutorial themes	Spanish tutorial themes
Environmental justice	Agricultural pesticide use and human health impacts in Guadeloupe and Martinique	Cancer villages; environment activism and activists in China	Asbestos in automotive friction products; environmental compliance of artisanal gold mining in Colombia
Climate change and wildfires	The impact of climate change on the wine industry in France; climate change and the Congo River Basin	Climate change and flood risks in the Yangtze River Delta; the impact of climate change and food security in China	Protection of Colombian paramos; Amazon Jungle: political fires?
Hydropolitics	Alternative energy— HydroQuébec and hydropower; nuclear energy in France	Three Gorges Dam and hydropower in China; drinking water crisis in China	Strategic investment for Paraguay's sustainable development; Costa Rica's adoption of renewable resources
Biodiversity	Biodiversity in the Democratic Republic of Congo and Cameroon	Panda conservation and biodiversity in Southwest China; biodiversity conservation and ecological tourism in ethnic minority populated areas	Latin America as a biodiversity hotspot; ecotourism in the Galapagos Island; collective local payments for ecosystem services in Mexico

and led the resulting discussion. In preparing a session on nuclear energy in France, for example, students analyzed the language and messaging of both pro-nuclear associations and anti-nuclear activists. This work also highlighted the 'wickedness' of nuclear energy in France, as the students emphasized environmental justice concerns at work in current discussions of the impact of nuclear waste storage and debated solutions with their impact on ecosystems and local communities in mind. These projects allowed students to make concrete connections between the content of the CLAC tutorial and their academic interests, with many students choosing to approach a 'wicked problem' that intersected with major fields of study or even prior internship experience.

For instance, one student's internship work for an organization devoted to endangered species awareness led her to develop a case study project tracing the links between deforestation, biodiversity, and zoonotic disease.

The second project asked students to identify an environmental problem relevant to a specific community in the Mandarin-, Spanish-, or French-speaking world, this time in order to create a two-part public service campaign to address the problem. Work began midway through the semester with an in-class examination of authentic public service announcements (PSAs), paying particular attention to the form and linguistic features and conventions of the genre. Students then researched their topic and workshopped their initial ideas and messaging with feedback from their instructor and peers and, where possible, from members of the target community. The final results, consisting of both a print PSA poster and a short audio or video spot, were presented first within students' respective CLAC courses and then to students in the other two CLAC tutorials during a culminating joint session. Students also shared their final work, accompanied by English transcription and explanatory text, with students enrolled in the ENVIRON 201 course.

Students gravitated toward final project topics that resonated strongly with previous experiences in their chosen region, or academic or personal interest. They reported enjoying the opportunity to collaborate with classmates and to take an active role in determining the content covered in their respective tutorials. As they prepared their final PSAs, they weighed considerations of audience, tone, platform, and the cultural relevance of their messaging alongside questions of how to communicate complex environmental concerns in a way that is broadly accessible. Presenting their work to the other CLAC sections allowed them to articulate the culturally- and linguistically-specific elements of their environmental PSAs, and to explore the ways in which the sustainability concerns referenced in their work paralleled, or diverged from, those from other parts of the world.

Community-Engaged Learning

In addition to listening to other cultures through engaging with diverse perspectives and voices in the course materials, opportunities were created for students to engage directly with these voices. It was also crucial to create space to center the professional expertise and on-the-ground experiences of experts, especially in domains where instructors' own content knowledge was lacking. To that end, each CLAC course included several visits, conducted in the target language, from invited guest speakers who underscored the tangible connections that exist between the local community of our university and the broader, global communities. To prepare for these visits, students first worked with relevant materials proposed by the guest speakers in order to acquire a basic familiarity with the topic at hand. Each session began with a brief presentation from the guest speaker, followed by a period of general Q&A and discussion.

The online nature of our Fall 2020 semester facilitated our work with guest speakers, some of whom joined our classes from out-of-state or abroad. In the Mandarin CLAC class, for example, synchronous online class sessions welcomed guest lectures from a research scientist in water security from NASA's Goddard Space Flight Center and an environmental scientist based at Duke Kunshan University in China. Other more local connections foregrounded the links between Durham and sustainability initiatives in Francophone Africa and Latin America. French CLAC students learned from the extensive on-the-ground experience shared by the founder of an NGO based in Durham that supports environmental and social justice issues in his home region of the Congo Basin, where the connection of the economic, environmental, and ethical pillars of sustainability, as well as the complex interactions of linguistic and cultural diversity, play out on both a local and global scale. The Spanish CLAC tutorial incorporated a series of invited talks highlighting the work of female environmental scientists on campus working with communities across Latin America to solve transcendental environmental management problems, such as rural community engagement in the renegotiation of Paraguay's Itaipu dam or ecosystem payments in Mexico, via community-based solutions. Students' community engagement also drew upon the connections forged via a multi-year collaboration with the Fundación Ayuda por Colombia, a Columbian NGO that serves vulnerable children and adolescents. At the heart of this collaboration is an exchange with a small group of children at the Fundación. To initiate the exchange, each child sent in a question related to the environment (for example, 'What is the source of the water we drink?') and was paired with one CLAC student. The instructor worked with students to connect the questions with one of the tutorial's modules and to identify the best way to answer the question, taking into consideration the children's ages and the particularities of their local environment. The students then recorded a 5-minute video response to the question, using compelling images, simplified Spanish terminology, and age-appropriate explanations that invited the children to take an active stand in responding to environmental problems (Virguez, 2021).

Student Perceptions of the CLAC Tutorials

Though the CLAC initiative has existed at the authors' university for a number of years, including two previous iterations of the Spanish 'Voices in the Environment' tutorial, Fall 2020 marked the first semester in which this curricular model included sustainability-related tutorials across multiple languages and in conjunction with the ENVIRON 201 course. The authors' approach to this multilingual interdisciplinary collaboration was therefore exploratory in nature, designed to gauge student interest in sustainability-related coursework in the target language, to understand students' perceptions of their own linguistic and content-knowledge gains, and to assess students' global learning. To that end, the authors, in collaboration with faculty from other universities within the CLAC Consortium, developed brief pre- and post-surveys based on the AAC&U's Global Learning VALUE rubrics. In total, 23 students enrolled in the three CLAC tutorials (seven in the Chinese, six in French, and ten in Spanish), and while the sample size was small, their feedback, gathered via anonymous post-semester surveys, provided valuable insights to inform next steps for future iterations of these offerings.

An initial survey administered during the first week of the semester shed light on students' motivation for enrolling in a tutorial. Responses revealed a heterogeneous group of students. A majority of respondents were either currently taking or had previously taken an Environmental Studies course, but only three of the 15 total respondents were pursuing an Environmental Studies major. Indeed, a majority indicated that it was the language-based, rather than the discipline-specific, content that was their primary motivator for enrolling in a CLAC tutorial. Thus, as the semester concluded, we sought to better understand students' perceptions of the impact these tutorials had on their learning. How did the opportunity to continue their study of the target language and culture through the lens of sustainability contribute to their perceptions of their developing language proficiency? How did this work impact their understanding of sustainability topics both broadly and in culturally- and linguistically-specific contexts? Finally, how did they perceive the impact their semester-long exploration of language and sustainability had on their ability to engage critically with diverse perspectives functioning within complex, interrelated systems? To that end, students were invited to complete an anonymous survey that asked them to reflect upon their experience in the CLAC tutorials. Though the survey size was again small—12 of the 23 students responded—when considered holistically, it provides useful insight into student perceptions of this pilot initiative and a fruitful basis from which to consider future language and sustainability work.

Student Perceived Gains in the Target Language

The development of interpretive, interpersonal, and presentational language skills in reading, speaking, and writing is a crucial component of the discourse competence that is at the heart of CLAC pedagogies, as well as the interpersonal competence that is foundational to sustainability education. As mentioned above, the language-based component of the CLAC tutorials was the driving factor in students' motivation for enrolling. Thus, to gauge students' impressions of the impact the CLAC tutorials had on their language proficiency, students were first asked to reflect upon their perceived growth in these modes of communication.[3] The majority of respondents reported either substantial or moderate perceived growth in using the L2 across all three modes, with the most significant gains in presentational speaking, presentational writing, and interpersonal writing skills (see Table 11.2). These gains are notable, especially when we consider that tutorials met only 14 times over the course of the semester.

The goal of this study was to gain a holistic understanding of overall gains across all tutorials, so we did not intend to break results down by language. Still, the particular language of the tutorial may help explain why a quarter of respondents reported little perceived growth in interpersonal speaking, interpretive listening, and interpretive reading skills. Two of the three students who responded in this manner were enrolled in the Chinese tutorial, a language in which it takes more time to make substantial linguistic gains, especially when compared to French or Spanish. While some students did not perceive significant gains in their language proficiency, they did note a commitment to learning languages and to incorporating them in their fields of interest. Two students commented:

> This course reaffirmed my commitment to strengthening my Chinese skills and to look for jobs in the Chinese clean energy space.

Table 11.2 Student perceived gains in target language

#	Question How much growth do you think you've made in the following areas?	Substantial % N		Moderate % N		A little % N		Very little to none % N		Total N
1	I can have a conversation with a native speaker about course content (*interpersonal speaking*)	33.33	4	41.67	5	25.00	3	0.00	0	12
2	I can deliver an oral presentation in class about course content (*presentational speaking*)	66.67	8	33.33	4	0.00	0	0.00	0	12
3	I can listen to a presentation or watch a video about course content without subtitles (*interpretive listening*)	33.33	4	41.67	5	25.00	3	0.00	0	12
4	I can read and comprehend a newspaper article or short article about course content (*interpretive reading*)	16.67	2	58.33	7	25.00	3	0.00	0	12
5	I can write a report or short paper about course content (*presentational writing*)	41.67	5	41.67	5	8.33	1	8.33	1	12
6	I can exchange ideas and opinions about topics covered in the course via email or text (*interpretive writing*)	50.00	6	41.67	5	8.33	1	0.00	0	12

This course has broken me out of my comfort zone with the language that I'm studying by forcing me to apply it in a way that I never had before. It has opened my eyes to the possibility of integrating my interest in language with my love and passion for the environment.

Students' Perceived Gains in the Course Content

As previously mentioned, the language tutorials were designed to complement the core ENVIRON 201 course, though enrollment in ENVIRON 201 was not required. During the Fall 2020 semester, only two students enrolled concurrently in both ENVIRON 201 and a corresponding CLAC tutorial. Nevertheless, since aligning the tutorials' content with that of the ENVIRON 201 core course was a central curricular design aspect, student feedback on the perceived gains in their understanding of sustainability-related content was vital to the authors' understanding of the success of this pilot phase of this project. To that end, students were asked to indicate their perception of the ways in which their work in a CLAC tutorial enhanced their understanding of core sustainability concepts within the context of the target cultures (see Table 11.3). Nearly all students thought that working in/through another language improved "a lot" or "a great deal" their ability to identify major environmental challenges related to the culture/s of their CLAC tutorials and helped them develop a culturally-specific understanding of the effect of environmental problems on local populations. Similarly, all student respondents felt that the CLAC tutorials improved their understanding of environmental studies content. More importantly, they deepened their understanding of how language and culture impact people's views about global environment issues. Through case studies, project-based inquiry, and community engagement,

Table 11.3 Student perceived gains in discipline-specific content

#	Question Did working in/through another language…	A great deal % N	A lot % N	A little % N	Not at all % N	Total N
1	Improve your understanding of the content in the core discipline?	41.67 5	50.00 6	8.33 1	0.00 0	12
2	Improve your ability to identify major environmental challenges related to the culture/s of your CLAC course?	66.67 8	33.33 4	0.00 0	0.00 0	12
3	Help you develop a culturally-specific understanding of the effect of environmental problems on local populations?	58.33 7	41.67 5	0.00 0	0.00 0	12

students identified environmental challenges in a specific culture and ana-lyzed how language and culture play a role in environmental issues in other cultures. As one respondent elaborated:

> The CLAC course opened my eyes to how environmental studies share similarities, but also unique differences across cultures that affect the knowledge, research, and response to addressing key environmental challenges. It also emphasized the importance of considering the role language plays in prioritizing and communicating certain environmen-tal issues—and which communities are most able to address these. This amplified my interest in study abroad that combines community-based research with studying French to gain new perspectives from different people about how the environment affects their well-being.

Students' Perceived Gains in Global Learning Skills

Over the course of the semester, the aforementioned AAC&U's Global Learn-ing skills informed the instructors' approaches to working with the tutorials' content. Students had been introduced to the ways in which the six targeted global learning skills—(1) global self-awareness, (2) perspective taking, (3) cul-tural diversity, (4) personal and social responsibility, (5) understanding global systems, and (6) applying knowledge to contemporary global contexts—tied into the thematic units at the beginning of the semester, and they received feedback related to these skills on their midterm and final projects. Thus, to understand students' perceived gains related to these skills at the end of the semester, the authors returned the AAC&U's Global Learning VALUE rubric framework (AAC&U, 2014; Rhodes, 2010). The survey question indicated in Table 11.4 framed the six global learning skills as can-do statements.

All respondents believed that they made some growth in all global learning skills, though the extent of this growth varied noticeably. For example, while more than half of respondents perceived substantial growth in their ability to learn respectfully about cultural diversity (question 3), three-fourths of stu-dents indicated little to moderate growth in their perspective-taking ability (question 2), their understanding of the relationships between local and global communities (question 1), and their ability to apply knowledge and skills in real life situations (question 6). These results may point to, on the one hand, the difficulty inherent in gauging one's own progress in conceptual skills like perspective-taking, which are harder to quantify than language proficiency and content knowledge. They also indicate that, in future iterations of these tutorials, students need more opportunities to intentionally consider the local and global intersections of these 'wicked problems,' perhaps via more struc-tured reflection, expanded interactions with guest speakers, or even expanded opportunities for community engagement.

Additionally, while even modest perceived gains suggest that CLAC courses can have the potential to develop students' global learning skills from

Table 11.4 Student perceived gains in global learning skills

#	Question How much growth do you think you've made in the following areas?	Substantial %	N	Moderate %	N	A little %	N	Very little to none %	N	Total N
1	I can understand the interrelationships between myself, my local community, and global communities (*global learning skill: global self-awareness*)	33.33	4	50.00	6	16.67	2	0.00	0	12
2	I can engage and learn from perspectives and experiences different from mine and understand how my place in the world both informs and limits my knowledge (*global learning skill: perspective taking*)	25.00	3	58.33	7	16.67	2	0.00	0	12
3	I can stay curious to learn respectfully about the cultural diversity of other people and recognize the origins and influences of my cultural heritage (*global learning skill: cultural diversity*)	58.33	7	33.33	4	8.33	1	0.00	0	12
4	I can recognize my responsibilities to society—locally, nationally, and globally. (*global learning skill: personal and social responsibility*)	50.00	6	33.33	4	16.67	2	0.00	0	12
5	I can understand complex and overlapping worldwide systems—natural systems and human systems. (*global learning skill: understanding global systems*)	50.00	6	41.67	5	8.33	1	0.00	0	12
6	I can apply knowledge and skills gained to real-life problem-solving, alone and with others. (*global learning skill: applying knowledge to contemporary global contexts*)	25.00	3	58.33	7	16.67	2	0.00	0	12

Can-do statements adapted from the AAC&U's (2014) Global Learning VALUE rubric.

both a language and sustainability education perspective by internationalizing the post-secondary curriculum (Bettencourt, 2011; Klee, 2009; Reisinger et al., 2015), a single CLAC tutorial should not be considered an end point, but rather as one component of what could become additional language-and-content courses, or even curricular clusters. These expanded course offerings could provide students further opportunities to explore sustainability problems from a multilingual perspective.

In response to a final question that asked students to reflect upon the impact their CLAC tutorial had on their future course selection, a large majority of students reported that the tutorial made them very or somewhat likely to take a content course in the core discipline (83.3%), to enroll in an additional language course (100%), to participate in a study abroad program or a service-learning course in the future (83.3%), and even to major or minor in the language (91.6%). Furthermore, 92% of students said that they would be very likely to recommend a/this CLAC course to another student. These responses suggest that in addition to providing students unique opportunities to explore sustainability topics via the lens of other languages and cultures, CLAC tutorials such as these have the potential to increase student interest in continuing their study of the language itself by presenting students with a curricular option that falls outside the literature-focused courses that have traditionally made up advanced-level language study.

CLAC Courses: Challenges and Opportunities

The design and implementation of this pilot cluster of language tutorials was immensely rewarding but was not without its challenges. One initial challenge, related to the community engagement component of the tutorials, arose due to the online environment made necessary by the COVID-19 pandemic. Given both the need to socially distance and the fact that students were in multiple locations and time zones, it was not possible to implement the kinds of community-based activities (such as volunteer work in schools, field trips, or activities related to the local environment) that would be a natural fit for this type of course. While both ENVIRON 201 and CLAC students were able to come together on one occasion to create an environmentally-themed community mural, the majority of our engagement had to be managed in an online environment via weekly Zoom discussions and guest speaker interactions.

Another initial challenge was the Chinese and French instructors' lack of expertise in environmental studies. To address this, they worked closely with the Spanish environmental scientist, meeting frequently throughout the semester to exchange ideas about course material selections and curricular activities and to reflect upon teaching strategies. They also met regularly with the instructor of ENVIRON 201, herself an environmental scientist specializing in post-secondary sustainability education, to align course pacing, brainstorm effective approaches to the content, and suggest complementary approaches to students' project-based activities. The interventions of invited speakers in each of the tutorials provided further opportunities for students and instructors alike to benefit from expertise in diverse fields within sustainability. This collaborative response to perceived gaps in content knowledge reflects many of the current best practices in the field of sustainability education and the fundamentally transdisciplinary nature of sustainability more generally (Evans, 2015).

Indeed, interdepartmental collaboration and enthusiasm is key to the successful implementation of a project such as this. For practitioners interested in creating similar programs, the authors recommend identifying potential interdepartmental partnerships that align with the university's strategic goals. While all disciplines can benefit from expanded international perspectives in their core courses, successful CLAC initiatives draw on existing faculty expertise and student academic interests. Fruitful potential partnerships can arise, for example, by identifying faculty who are conducting research in international arenas, and the disciplines with which students most frequently pair a language double major in (e.g. at the authors' institution, majors such as Global Health or Public Policy are often paired with language majors). Students and faculty both gain synergistic benefits from these connections. The teamwork element of course creation also enhances faculty collaboration and professional development. Finally, establishing connections with communities should play a central role in course design and implementation, in order to address the 'glocal' and social dimensions of sustainability.

Conclusion

The development of new programmatic content exploring sustainability-related problems should be examined from multidisciplinary and multicultural perspectives. As demonstrated through the CLAC tutorials, the opportunity to offer a holistic, multilingual understanding of global challenges represents a valuable contribution to students' engagement with and increased proficiency in both linguistic and sustainability content. Teaching them how to navigate between multiple threads of knowledge and then synthesize these threads around a central idea fosters skills in identifying core trends and common patterns while establishing connections within them. By exploring environmental wicked problems from an interdisciplinary and culturally- and linguistically-informed perspective, students are able to consider new connections between sustainability-related challenges, their courses of study, and their study of language.

Notes

1 The Association of American Colleges & Universities' VALUE rubrics are a collection of rubrics tailored to 16 essential learning outcomes. These rubrics—which address intellectual and practical skills, personal and social responsibility, and integrative and applied learning—are intended to assess student learning cumulatively over the course of an entire college career. They

are also an effective tool for defining learning objectives at the course level. Defined as "engagement with complex, interdependent global systems... and their implications for people's lives and the earth's sustainability," the benchmarks outlined in the Global Learning rubric were a logical choice for this pilot phase of the project given their significant overlaps with CLAC's four methodological pillars (AAC&U, 2014).

2   While instructors endeavored to incorporate published case studies whenever possible, in instances where a published case study was either not available or inaccessible for students' proficiency level, instructor and students worked together to co-curate working case studies based on authentic materials representing a variety of perspectives on a given issue.

3   For more information on the interpretive, interpersonal, and presentational modes of communication, as well as proficiency-level performance descriptors for each mode, see the American Council on the Teaching of Foreign Language's (2012) Performance descriptors for language learners.

References

Association of American Colleges and Universities (AAC&U). (2014). *Global Learning VALUE rubric*. Retrieved from https://www.aacu.org/value/rubrics/global

Beckett, G. H., Slater, T., & Mohan, B. (2019). Philosophical foundation, theoretical approaches, and gaps in the literature. In H. G. Beckett & T. Slater (Eds.), *Global perspectives on project-based language learning, teaching and assessment* (pp. 3–22). New York: Routledge.

Bettencourt, M. (2011). Languages across the Curriculum: A response to internationalization in foreign language education. *Multicultural Education, 19*(1), 55–58.

CLAC Consortium. (2019, December 6). Retrieved from https://clacconsortium.org/

Clark, C. (2016). Collective action competence: An asset to campus sustainability. *International Journal of Sustainability in Higher Education, 17*(4), 559–578.

DeFries, R., & Nagendra, H. (2017). Ecosystem management as a wicked problem. *Science, 356*(6335), 265–270.

Evans, T. (2015). Transdisciplinary collaborations for sustainability education: Institutional and intragroup challenges and opportunities. *Policy Futures in Education, 13*(1), 70–96.

Geisler, M., Kramsch, C., McGinnis, S., Patrikis, P., Pratt, M., Ryding, K., & Saussy, H. (2007). Foreign languages and higher education: New structures for a changed world: MLA Ad Hoc Committee on Foreign Languages. *Profession*, 1, 234–245. Retrieved August 17, 2021, from http://www.jstor.org/stable/25595871

Hovland, K. (2014). *Global learning: Defining, designing, demonstrating*. American Association of Colleges and Universities. Retrieved April 1, 2021 from https://www.aacu.org/globallearning

Klee, C. A. (2009). Internationalization and foreign languages: The resurgence of interest in languages across the curriculum. *Modern Language Journal, 93*(4), 618–621.

Reisinger, D. (2018). Measuring the impact of the cultures and languages across the curriculum: New research directions. In D. Sorenson & C. Zimmer (Eds.), *Developing responsible global citizenship through cultures and languages across the curriculum (CLAC): Selected papers from the 2016 CLAC conference* (pp. 17–39).

Minneapolis: University of Minnesota, Center for Advanced Research on Language Acquisition.

Reisinger, D., Clifford, J., Whetten, K., & Deardorff, D. K. (2015). Cultures and languages across the curriculum in global health: Fostering empathy to develop intercultural competency. In W. Green & C. Whitsed (Eds.), *Critical perspectives on internationalising the curriculum in disciplines: Reflective narrative accounts from business, education and health* (pp. 261–274). Rotterdam: Sense Publishers.

Rhodes, T. (2010). *Assessing outcomes and improving achievement: Tips and tools for using rubrics*. Washington, DC: Association of American Colleges and Universities.

Rittel, H. (1973). Dilemmas in a general theory of planning. *Policy Sciences, 4*(2), 155–169.

Virguez, E. (2021). Embracing the value of cultural wealth from underrepresented groups. In K. L. Armstrong, L. A. Genova, J. W. Greenlee, & D. S. Samuel (Eds.), *Teaching gradually: Practical pedagogy and classroom strategies for graduate students by graduate students* (pp. 190–196). Sterling, Virginia: Stylus Publishing LLC.

Wiek, A., Withycombe, L. & Redman, C. L. (2011). Key competencies in sustainability: A reference framework for academic program development. *Sustainability Science 6*, 203–218.

Zilmer, C. (2018). Measuring the impact of the cultures and languages across the curriculum: New research directions. In D. Sorenson & C. Zimmer (Eds.), *Developing responsible global citizenship through cultures and languages across the curriculum (CLAC): Selected papers from the 2016 CLAC conference* (pp. 4–15). Minneapolis: University of Minnesota, Center for Advanced Research on Language Acquisition.

12 Translanguaging in Language and Area-Studies Curriculum

A Japanese FLAC Course of *Minamata* and *Fukushima* in Environmental Humanities

Nobuko Chikamatsu

Introduction

On March 11, 2011, Tohoku, the northern coastal region of Japan, was hit by the worst earthquake ever recorded in modern times. The following day, the Fukushima Daiichi Nuclear Power Plant exploded, filling the air and water with radioactive contamination. Almost a decade later, the recovery process is slow and proceeds without a clear vision of a new energy policy for Japan. Almost 50 years earlier, on May 1, 1956, a doctor in a small town in Japan's southern coast first reported the appearance of severe and seemingly unexplainable symptoms in the central nervous system of a five-year-old girl. Undetected by sight or smell, poisonous mercury had drained into the bay from a local chemical factory, decimating the health and livelihoods of the town's residents. The resulting illness became known as Minamata Disease. The lawsuits and injuries from this incident have continued to this day. These two industrial disasters, Fukushima and Minamata, remain unhealed scars in Japan's history, juxtaposed against the nation's dependence on heavy industrialization to maintain the comfort and convenience of its modern economy and lifestyle. Learning about these environmental and industrial tragedies and how to prevent them is valuable for any person, but especially for advanced learners of Japanese who have gained an appreciation of the richness of Japanese culture and society through their years of language learning. Nonetheless, negative legacies of Japanese history, 負の歴史, *fu no rekishi*, are often not mentioned in language curricula or textbooks (Chikamatsu, 2019; Kubota, 2016). Even at advanced levels, the complexity of these topics is often viewed as beyond learners' language proficiency, since Japanese is a Group IV language, the most challenging category for English native speakers to acquire[1] (Liskin-Gasparro, 1982).

This chapter illustrates how these environmental and industrialization issues can be successfully incorporated into Japanese language curricula using the Foreign Language Across the Curriculum framework (FLAC), which

DOI: 10.4324/9781003080183-16

offers learners the opportunity to develop L2 proficiency by taking course content offered in the L2, sometimes in association with another course with similar content in the L1 (Klee, 2009), while acquiring a cross-disciplinary perspective on the content. In particular, I describe a collaboration between Japanese language and area-studies faculties that focused on environmental ethics and resulted in two interdisciplinary courses that are taught concurrently. A full-credit, comparative ethics course is taught in English by an ethics specialist and covers global environmental issues, industrial diseases, and religious and ethical implications of industrialization. Concurrently, a half-credit language course is taught by a language specialist to advanced learners of Japanese with a focus on two case studies—the Minamata disease and the Fukushima nuclear disaster. This FLAC course includes two key learning activities—translating a Japanese picture book and a poem into English. Together, these two courses explore issues in *environmental humanities*, an interdisciplinary area of research that connects the approaches of the humanities to the environmental problems we face. These courses, taught in two languages, enhance reciprocal content and language learning.

Theoretical and Pedagogical Framework: Translanguaging in Japanese FLAC Courses

FLAC pedagogy is part of the broader Content-Based Instruction (CBI) framework, a language-learning approach where the target language serves as a vehicle to build academic content knowledge instead of the main object of study (Caldwell, 2001). CBI courses are developed based on subject matter rather than linguistic forms or functions (Chikamatsu, 2011, 2012; Rodgers, 2015; Stryker & Leaver, 1997). Thus, CBI courses can be implemented using the FLAC model, where a foreign language is taught in conjunction with a course in another discipline, such as history, culture, literature, etc.

In the FLAC framework, students can enhance language learning through content learning without explicit language instruction (Hanabusa, 2015; Kaufman, 2004). In the mid-2010s, FLAC courses were implemented within multiple language programs in the author's university. In spite of the raising numbers in Japanese advanced learners, which have enhanced the demand for CBI courses over the last decade (Chikamatsu, 2015; Douglas, 2017; Sato et al., 2015), the required increases in the number of students per classroom (due to universities' financial constrains) has made it challenging for less commonly learned languages to maintain full-credit advanced CBI courses (Ananth & Lyons, 2020). FLAC is viewed as a solution to this situation, allowing language programs to maintain advanced level Japanese instruction with smaller enrollments.

Choosing a theme for a FLAC course can be challenging due to today's diverse language learners of Japanese. While it can be discipline-specific, such as natural science, engineering, business, or law, it may also be meaningful to use a theme to explore subjects in humanities, which are universally important and critical for human activities (Okazaki, 2009). For this reason, the area of environmental humanities is an emerging multidisciplinary research field in higher education. Japan is no exception, as the discipline of Japanese Studies has included the relationship between nature and humankind as a central research and curricular theme, from ancient Shinto religion to contemporary apocalyptic science fiction. At this author's institution, the Fukushima nuclear disaster in 2011 had a significant impact: An environmental ethics course, taught by a Japanese native scholar, was revised to incorporate issues and challenges from 3.11; parallel to it, the first Japanese FLAC course was designed and first offered in 2015.

This 'marriage' between disciplinary coursework in environmental ethics—in the L1—and advanced language work—in the L2—offers numerous benefits for advanced language pedagogical work in Japanese. Recently, research into transdisciplinary frameworks for second/foreign language learning has set the stage to reevaluate the multifaceted nature of language learning and teaching in our diverse modern world (Clark, 2016; Hall, 2019; The Douglas Fir Group, 2016). Learners utilize and develop semiotic repertoires through fluid and flexible multilingual and multimodal resources. Consequently, the "learning environment should include purposeful activities that emphasize processes of experiencing, conceptualizing, analyzing and applying" (Hall, 2019, p. 37), and instructors should see that "teaching a second language is about creating increasingly complex meaning-making contexts in which students can use their cognitive and other capabilities to expand on the repertoires they already have" (Hall, 2019, p. 36).

However, implementing these principles is not an easy task, in part because Japanese is challenging, as it is a Group IV language. With linguistic features that are not shared with Romance languages, from logographic orthography to structural-stylistic variations, the range of advanced courses offered in Japanese is somewhat limited. Further, learner proficiency may not be sufficient to deal with original publications in Japanese, such as novels or scholarly articles. As a result, civilization and literature courses commonly offered in the target language in Group I or II languages (e.g. Spanish or French) are usually taught in English in North America by literature or area-studies specialists. The FLAC framework offers a unique opportunity to overcome these issues through the concurrent learning of the same theme in English and Japanese, enabling learners to discuss complex topics in the L1 but also the L2: Japanese.

A strong argument for using the FLAC instructional model for Group IV language pedagogy can be made based on general cognitive processes of learning (Anderson & Krathwohl, 2001). Japanese language instruction often places significant emphasis and time on building lexical knowledge or text comprehension with lengthy vocabulary lists and reading comprehension questions or lower-level thinking activities. Time often runs out in class before reaching higher-level thinking tasks, namely analysis, evaluation, and creation of ideas, as defined in Bloom's taxonomy (Anderson & Krathwohl, 2001). Under the FLAC framework, however, less cognitively demanding tasks can be completed in the 'stronger' language, the L1. This allows instructors to devote class time to discussion, thereby helping learners express their thoughts and opinions in the L2.

The field of second language learning, or Second Language Acquisition (SLA), also offers us support for FLAC pedagogy. In particular, the 'translanguaging' that occurs under this FLAC model, where students discuss themes going back and forth between their L1 and their L2, is said to promote L2 development (García & Wei, 2014; García & Kano, 2014). The concept of translanguaging has its roots in dynamic bilingualism, "which refers to the multiple language interactions and other linguistic interrelationships that take place on different scales and spaces among multilingual speakers" (García & Wei, 2014, p. 51). With increasing cross-border immigration and virtual communications, a bilingual individual manages two (or more) languages interchangeably and simultaneously for different needs and occasions and activates bilingual resources for communication as one integrated repertoire rather than two sets of separate languages (Leung & Valdes, 2019). This process is called translanguaging, consisting of "accessing different linguistic features or various modes of what are described as autonomous languages, in order to maximize communicative potential" (García, 2009, p. 140).

Kanō (2016) points out the benefits of using translanguaging during classroom instruction, primarily when the stronger language (L1) is used to develop the weaker language (L2). For instance, a learner's L1 is used for reading comprehension of an L1 text, while the learner's L2 is used to write an essay. Kanō found that such fluid usage of bilingual resources can (i) strengthen L2 comprehension with more comprehensible input, (ii) help learners think and discuss more deeply and critically, and (iii) develop literacy in both languages. Kanō also found pedagogical advantages, as translanguaging can make it possible to teach different proficiency levels together in one class, and allow learners who could not express themselves in L2 to have their own voice, interact, get feedback, and comprehend in their L1. This enables 'weaker' L2 students to blossom in a translation course or other translanguaging pedagogy setting.

Kumagai and Kono (2018) applied translanguaging learning in FLAC Japanese courses: a literature course taught in English (L1), and an advanced language course taught in Japanese (L2), offered concurrently by two instructors, where the content's focus was on ethnic minorities in Japan. The advanced learners of Japanese, who were enrolled in both courses, "had access to materials in both English and Japanese and could use their knowledge of each language to enhance their use and understanding of the other language" (p. 257). In the courses, translation was essential for developing critical literacy and integrating content and linguistic knowledge into language practice (Hasegawa, 2012; Someya, 2010). Students read both original and translated texts and produced their own translation work.

Using a similar theoretical foundation and pedagogical framework, a FLAC Japanese language course in conjunction with an environmental ethics course was developed that addressed crucial environmental sustainability content. The following sections offer an overview of these courses and a description of their curricular designs.

The FLAC Courses

The current Japanese FLAC course (JPN hereafter) has been offered three times since 2015 in the Modern Languages Department, paired with an Environmental Ethics course (REL hereafter) offered yearly in the Religious Studies department. While some course materials may vary year to year, these courses' overall content and goals remain constant, as described below.

Religious Studies Course (REL)

The REL course is entitled *Religion and Ethics: Industrial Diseases in Comparative Perspectives,* and it is taught in English by a native Japanese specialist. This is a full-credit ten-week, three-hour/week course that satisfies the Liberal Studies or Religious Studies major/minor requirement. Yearly course enrollment includes 30 to 40 students from all disciplines and years. It uses a primary textbook, *Toxic Archipelago: A History of Industrial Disease in Japan* (Walker, 2010), and over 30 articles and several videos are assigned to cover concepts and case studies from Japan and other countries. Key issues include modernity, industrialization, secularization, responsibility, the definition of (or lack thereof) nature, and environmental justice based upon discrimination, according to race, class, and gender.

Japanese FLAC Course (JPN)

The Japanese FLAC course is a half-credit, ten-week language course that meets once a week for 90 minutes, entitled *Industrial Diseases: Minamata and Fukushima.* It discusses two recent cases of industrial diseases in modern history: Minamata, caused by mercury poisoning in the 1950s, and Fukushima, afflicted by radiation in 2011. These topics are also discussed in the REL course. The goals of the JPN course are to (i) develop critical thinking skills for linguistic and literary analysis of Japanese textual and visual materials, (ii) examine environmental, socio-cultural, and gender issues in modern industrialization and destruction; and (iii) make a proposal to contribute to communities as a bilingual individual. The course is designed for advanced students at the fourth year or higher level of Japanese who are concurrently enrolled in the REL course. Three[2] Advanced Japanese students were enrolled in the JPN course each time it was offered between 2015 and 2019 (Year A, Year B, and Year C, hereafter). Table 12.1 outlines the contents of each course. The first half of the JPN course is devoted to the Minamata case and the second half to the Fukushima case. Authentic textual and visual materials used in class include a children's picture book, manga, anime, poetry, supplementary textbooks, an essay, parody songs, and puns. A homework worksheet is

Table 12.1 Course outline of Japanese FLAC and environmental ethics courses

Weeks	JPN topics and materials	REL topics
1	Introduction and keywords *Dialects and Maps* 方言と地図 (Izuka, 2009)	Why religion? How to study ethics? Religion and environmental discourse: Environmental disasters/Industrial diseases; Religion and nature in the year of climate change
2	Minamata: Picture book *A Minamata Tree* みなまたの木 (Mieda, 2011)	Religion/s in Japan: Japanese religions; Between religions and *Shūkyō*
3	Minamata: Manga *Blackjack* ブラックジャック (Tezuka, 1970/80s)	Religion and nature: In Catholicism; In Buddhism
4	Minamata: Anime *Nausicaä of the Valley of the Wind* 風の谷のナウシカ (Miyazaki, 1984)	Secularism, modernism, and industrial diseases: Minamata I; Science and disenchantment
5	Minamata: Literature *Paradise in the Sea of Sorrow* 苦海浄土 (Ishimure, 1964), [Term project proposal]	Intrinsic and instrumental value of environment: The agency of insects and chemicals
6	Fukushima: Education *Booklet of Radiation* 小学生のための放射線副読本 (Ministry of Education, 2013) & *Dr. Koide's Search for Anti-Nuclear Power* (Koide, 2012) 原発に反対しながら研究をつづける小出裕章さんのおはなし	Anthropocene, animism, and anime: Animism and art; Re-enchantment and anime
7	Fukushima: Essay (by a student) *I Still Buy Peaches* それでも僕は桃を買う (Ōnuma, 2013)	Nature and nationalism: Nature as propaganda; Nature and nationalism
8	Fukushima: Songs Parody songs (of anti-nuclear power)	Environmental justice: Gender inequality in industrial diseases; Minamata as peripheral
9	Fukushima: Puns *Cool Gags in Your Heart* いつも心にクールギャグを (Specter, 2011)	From Minamata to Fukushima: Spirituality in Minamata; Fukushima
10	Presentation Term project presentation	Post-Fukushima; Food after Fukushima, culinary nationalism

Events	Year A: On-campus Film Screening *Tohoku Tomo* 東北友 (Julian, 2013) Year B: *The Atomic Age IV Symposium* (2017) Year C: Chicago International Film Festival, *Shoplifters* 万引き家族 (Koreeda, 2018)

assigned weekly, and readings and/or visual materials based on classroom discussions. A term project is assigned for the second half of the course, which can be a research paper, book/film reviews, translation of Japanese originals, or any creative work, such as creating manga, songs, or short videos. On- and off-campus events are also incorporated, such as participating in the organization and screening of a 3.11 documentary film, participating in local academic conferences, and watching Japanese films covering social inequity at a local film festival.

Transforming Meaning and Developing Critical Literacy in the Japanese FLAC Course

To show how content and language learning interact across two languages in the REL course and JPN course, and the resulting benefits for L2 learners of Japanese, the following paragraphs examine two sample activities that illustrate (i) the productive meaning-making processes and their effects on content and linguistic analysis, and (ii) the development of critical literacy, creativity, and linguistic sensibility taking place during the environmental literary FLAC class.

Transformation of Meaning in Environmental Literacy: Minamata no Ki

L2 knowledge is a repertoire of diverse semiotic resources, and it is shaped through constant meaning-making practice in various contexts (Hall, 2019). This transformation of a semiotic resource is called *meaning potentials*, i.e., "conventionalized meanings that develop from their use" (p. 48). Meaning potentials are "context-sensitive, with particular meanings activated or motivated by the contexts in which they appear" (p. 48). Therefore, it is crucial to provide L2 learners with a complex, meaning-making context to build their own personalized meaning potentials and expand their semiotic resources. The following example displays how the transformation of meaning, *meaning potentials*, was exercised through the translation process of environmental literature work.

This activity was conducted with a picture book entitled みなまたの木, *Minamata no Ki*, (Mieda, 2011), tr. *The/A Minamata Tree*, which was introduced at the beginning of the JPN course, and later developed into a collaborative term project in Year A with three students. The tree, which narrates the book, depicts the story of a family of four—mother, father, big sister, and little sister. The family lived happily but often found themselves in financial instability due to the uncertainty of fishing, the father's profession before a chemical factory was built. The arrival of the factory polluted the village's water and caused the ultimate destruction of the town's nature and people. In the book, the little sister dies, followed by the father and mother. Even after years of hardship, the story ends with hope as the sole survivor of the family—the big sister, now an aged grandmother—visits the tree with her grandkids. While the book targets young readers, complex issues about socio-economic inequity, discrimination, and ecofeminism are subtly addressed. The book is written with unique linguistic and visual features, such as a vernacular dialect, nameless characters, and colored and colorless images. During Year A, in Week 2, the entire book was assigned along with a worksheet focusing on content and language analysis. Interestingly, *Minamata no Ki* was chosen again for the term project, not by design, but upon the students' request. In Week 6, all three students decided to translate *Minamata no Ki* into English as a group project. In Week 7, individual translations were submitted, and between Week 8 and Week 10, these translations were reviewed and revised collaboratively in and out of class.

The following passage is a scene on page 10 of the book that describes how industrialization began in the small fishing village. The current analysis focuses on the term 町, *machi*, which appears twice (indicated as 町[1] and 町[2]). The term is familiar to the students, commonly translated as a 'town' or 'city,' introduced in the first-year Japanese textbook and repeatedly used after that.[3]

このあたりの人に　　おちついたくらしをさせたいと　かんがえていた町[1]は
kono atari no hito ni　ochitsuita kurashi wo sasetai to　kangaeteita machi wa
here around p*　people p　calm　　life　　p　wanted to let p　had thought　MACHI p
大きな会社に　大きな工場を　つくってもらうことにした。
ōkina kaisha ni ōkina koujō wo tsukutte morau kotonishita.
large　　company p　large　　factory　p　create　had/asked decided to
町[2]のおおくの人は　　　その工場で　はたらくようになった。[4] (p. 10)
machi no ōku no hito wa sono kōjō de hataraku yōni natta.
MACHI　*p　many　p　people p that　factory　p　work　　started to*

*p=a particle which follows a noun, a verb, an adjective, or a sentence as a
suffix to indicate the meaning or function of the preceding word.

In the first individual drafts, submitted in Week 7, one student translated the passage as follows: *For these people who thought about living a calm life, a large company built a large factory. Many of the people in the town[2] began to work at the factory.* As expected, the second *machi*, 町[2], was translated as 'town' as in "the people in the town," while the first *machi*, 町[1], was missed. In the original Japanese, the first *machi* is indeed the agent of the action (the subject of the main verb), つくってもらうことにした, *tsukutte morau kotonishita*, "decided to have X build Y for Z." While this student's translation, "a large company built a large factory" is not entirely incorrect, the first *machi* was missed as the agent of decision-making. In class, we discussed what this first *machi* would really mean in the context. Then, the student group identified it as the force that decided to bring a large company into the town—in other words, the town's government or chamber of commerce.

In the second collaborative draft, submitted in Week 8, the first *machi* was translated to "the town's government" whereas the second remained "the town," as follows: *The town's government[1] brought a large company to the town for the people to have a more stable life. Many of the people in the town[2] began to work at the factory.* Then, in the classroom discussion, the students shifted focus to another term, 大きな, *ōkina*, 'big' used twice as in 大きな会社に大きな工場, *ōkina kai-sha ni ōkina kōjō*, "big company, big factory." A distinct contrast in the power structure was noted through the word *ōkina*, which could represent the power or authority, i.e., a big corporation from a big city, as opposed to the powerless local government or people. Accordingly, *machi* was revised once more in the third draft as follows: <u>*The local government*</u>[1] *brought a large company to the town for the villagers to have a more stable life. Many of the people in* <u>*the village*</u>[2] *began to work at the factory.* The students consciously changed "town" to "local" for the first *machi*, and "town" to "village" for the second in the final version. The second *machi* in 町の多くの人, *machi no ōku no hito*, turned to "villagers," i.e., mar-ginalized rural people suffering at the hands of the big corporation and the central government. The notion of *dichotomies*, such as between government and people, urban and rural, and rich and poor, is one of the main causes for environmental injustice discussed in the REL course. Thus, the single word *machi* transforms into "the local government" for one case and "villagers" for another. Such profound linguistic and literary analysis was only possible be-cause of a thorough understanding of the ethical theories and perspectives discussed in the concurrent REL course in English. *Machi* indeed appears many more times throughout the book, and each was carefully translated into different words hereafter, such as "village," "town," "city," "government," "people," and "villagers."[5]

Versatility and Creativity in Environmental Literary Translation: Kugaijōdo

Literary translation requires us to read a text critically and often read between the lines. Critical literacy helps readers to become aware of how linguistic information—such as lexicon, modality, and voice—mediate particular per-spectives in the given text (Koyama, 2016; Kumagai & Kono, 2018). Under-standing ambivalence, ambiguity, metaphor, connotation, and implication requires both linguistic knowledge and an understanding of pertinent cul-tural, historical, and social background information. This understanding can be built and enhanced through fluid translanguaging settings. The following activity was designed to develop learners' critical literacy through a poetry translation process with the theme of environmental justice.

The poem is an excerpt from a semi-fiction book, 苦海浄土 わが水俣病 (1964), *Kugaijōdo Waga Minamatabyō*, tr. *Paradise in the Sea of Sorrow, Our Mina-mata Disease*. The book was written by Michiko Ishimure in 1964 and trans-lated into English by Livia Monnet in 2003. It depicts the first 15 years of the Minamata Disease of the 1950s and mixes victims' testimonies, excerpts from the clinical and government reports, and the author's own powerful de-scriptions of human suffering and dignity. A linguistic contrast is presented between the Minamata dialect of the victims' voices and the formal, standard Japanese used by central governments and corporate executives. Since the original Japanese work is lengthy and challenging for the students, a short poem written in the victims' voice was chosen to be read in Week 5, at the end of the Minamata unit. Students were assigned a worksheet as a reading guide to prepare for class discussion. The worksheet contained questions, such as who wrote this poem, to whom, and why, or what specific linguistic features are included (such as dialect or a modal auxiliary verb) and why. Students were also asked to translate the poem into English in the worksheet, paying attention to emotion and tone.

銭は1銭*もいらん**。
zeni wa issen mo iran**
*money p** one sen* need no*
そのかわり、会社のえらか衆の、
Sono kawari　kaisha no eraka shū no
Instead　　　company p authoritative people p
上から順々に、水銀母液ば飲んでもらおう。
ue kara junjun ni suigin boeki ba nonde moraō
top from sequentially p mercury mother liquid p drink AUX
上から順々に、42人死んでもらう。
ue kara junjun ni 42 nin shinde morau
top from sequentially p 42 people die AUX
奥さんがたにも飲んでもらう。胎児性の生まれるように。
okusan gata ni mo nonde morau　taijisei no umareru youni
Madams p drink　　　　　　AUX fetus born will so that
そのあと順々に69人、水俣病になってもらう。
sonoato junjun ni 69 nin Minamata byō ni natte morau
after that sequentially 69 people Minamata disease p become AUX
あと100人ぐらい潜在患者になってもらう。
Ato 100 nin gurai senzaikanja ni natte morau
Then 100 people around potential patient become AUX
それでよか。
sore de yoka
that p good

*sen: an old currency's smallest unit
**p: a particle that follows a noun, verb, adjective, or sentence as a suffix to indicate the meaning or function.
_____: dialectal expressions used in the Minamata area
------: auxiliary suffix verb (AUX), te morau, which implies the speaker's appreciation for someone else's action of the preceding verb

After the discussion of the worksheet, each student shared their own translation and critiqued each other's. Then they were asked to create one polished translation collaboratively. The following translations were created in Year A, B, and C, as the final version at the end of the class.

Year A
I got no need for even a penny.
'druther those high-and-mighty company folk,
One by one, drink the mercury moonshine for me.
One by one, 42'll die for me.
The Missus will drink for me,
to pass down the sickness.
And then 69'll get the decease for me.
Another 100 ailing will get recognition for me.
That'll do.

Year B
We don't need your money.
Instead, the people from the corporate head
Should drink, one by one, the liquid mercury
One by one, 42 people should die.
So their children have it too, their wives should also drink.
Then, one by one, 69 more people should get the disease.
In addition, 100 more people should become patients.
Is that okay with you?

Year C
We don't need a single cent.
In exchange, from the top of the company's admirable masses in turn,
One by one, do yourself the favor of drinking the mercury
One by one, watch 42 of your dies.
Have your wives drink it too and birth defected babies.
After that, one by one, 69 more people contract Minamata Disease
And then allow 100 more people to fear that they're next.
That's what we want.

While the overall meaning is similar across the three translations, each contained unique wordings and perspectives, which resulted in different emotions and tones (see Table 12.2). For instance, because the original Japanese poem contains dialectal expressions, such as いらん, *iran,* "no need," えらか, *eraka,* "authoritative," andよか, *yoka,* "good," Year A students translated the poem with unconventional words, such as *'druther, 42'll* or *moonshine,* keeping the vernacular tone, as to suggest the victims' marginalized position in society.

The modal beneficiary auxiliary verb suffix, てもらう, *te morau,* which is conventionally used to describe one's action in the speaker's favor, was used repeatedly with cynicism for the company's violent actions toward the victims. Year A translated it as *for me* to keep the original cynical tone, Year B as *should* to state more evident accusation and blame the company's responsibility, and Year C as imperative forms, such as *do yourself* or *watch* to demand the company do as they did to the victims, as revenge.

Table 12.2 Comparison of collaborative translations by year

	Year A	Year B	Year C
Vernacular (Dialectal)	Included '*druther; high-and-mighty; moonshine; 42'll; Misses,* etc.	Not included	Somewhat included contracted form, don't; they're; that's
Modal verb suffix (Receiving-favor)	*for me*	*should*	Imperatives, e.g., *the favor of doing ~; have ~*
Pronouns (1st to 3rd persons)	1st: *I* 2nd: none 3rd: none	1st: *we* 2nd: *you* 3rd: *they*	1st: *we* 2nd: *you* 3rd: none
Approach: Literal vs. meaning-based • 胎児性の生まれる • 水俣病になる	More meaning-based • pass down the sickness • get the disease	More meaning-based • their children have it too • get the disease	More literal • birth defected babies • contract Minamata Disease
Ending (*italic*) • それでよか	That'll do.	Is that okay with you?	That's what we want.

Differences were also observed in the choice of pronouns. Japanese has a pro-drop, non-inflection feature where the subject or object of a sentence may be dropped from the surface structure without any trace of person, number, or gender in verb inflection. Therefore, who or what is being referred to must be inferred from the context. Consequently, the choice of pronouns (for translation into English) must be made based on the reader's interpretation of the original text. Only the first person (*I* or *me*) was used in Year A's translation to make it sound personal, as if the students tried to stand in the victim's position and express their compassion. Year C students, instead, used both first and second person pronouns to reiterate the power dichotomy between victims (*we*) and victimizers (*you*). In turn, Year B students used all first, second, and third pronouns, as the translation starts with *we* and *you* in the first line, then switches to the third person, *they*, in *their wives* and *their children* of the fifth and sixth lines (c.f. *your* in Year C's), and ends with *you* in the last sentence. This switch would be adequate to present the victim's anger towards two different parties: First, talking in front of someone representing the company, such as lawyers or executives (*you*), and then confronting those who drained mercury at the factory (*they*).

The last short, powerful sentence, それでよか, *sore de yoka,* concludes with a specific tone and emotion in each translation: "That'll do." in Year A for the victims' hopeless or powerless feeling, "Is that okay with you?" in Year B for their accusation to demand the company's responsibility, and "That's what we want" in Year C to articulate their rights and determination to fight for it.

Thus, students could read the original work critically and interpret the discrimination, inequality, power structures, social hierarchy, and dichotomies hidden between the lines. Using translanguaging, they were able to feel the victims' desperation, anger, and yearning for justice, then they recreated the poem in another language with their own unique selection of words. Such versatility of self-expressions was supported possibly by the thorough content analysis done in the REL course in the student's stronger L1 language, English.

As mentioned above, environmental literary works are instilled with the dichotomy of institutional power and subaltern. The authors of such works use linguistic devices, consciously or subconsciously, to express such perspectives. The translation process requires us to decode such devices consciously by moving back and forth between two languages. This blurs the boundaries between the two languages and develops semiotic repertoires in a learner through fluid and flexible multilingual and multimodal resources. Thus, translation is one essential facet of translanguaging and serves as a key tool in second/foreign language learning.

Discussion

Humankind's relationship with the natural environment has been a common theme in Japanese language classrooms, from traditional religious rituals to modern day garbage disposal[6] (Oka, 2009). Worship and respect toward nature were once an essential part of everyday life in Japan. However, Japan's modernization and industrialization have steadily reshaped the country's mindset and attitude toward the environment. Such shifts reflect on our language usage in literary work or daily communication. Chikamatsu (2019) claimed that the 3.11 disaster changed the usage of unfamiliar lexical terms, such as *mizō* 未曾有 ('unprecedented') and *sōteigai* 想定外 ('beyond imagination'), which suddenly became common expressions, as if ordinary words were inadequate to express the magnitude of damage and shock of 3.11. As time plays an essential role in transforming an individual's meaning-making semiotic systems, fluid usage of two or more languages, i.e., translanguaging, also helps dynamic bilinguals develop new ways of capturing meanings in today's digitally or physically interactive global society.

The fluid usage of English and Japanese in the two concurrent courses of environmental ethics and literature, the REL course and the JPN course, respectively, enabled learners of Japanese to discuss environmental justice and build learners' autonomy in versatile meaning-making processes across these two languages. Such reciprocal translanguaging learning was also received positively by the students. By examining end-of-semester surveys, we are able to see students' overall positive perceptions and attitudes toward concurrent content-language learning. Two aspects stand out:

i   students underscore the benefits of this form of 'bidirectional' or interdisciplinary learning:

   "(The JPN course) provided further context to material summarized in REL205."

"Content-wise, I really did benefit, especially in the latter half of REL205, the things we learned in JPN395 gave me a deeper perspective of things discussed in REL205."

ii   they report increased confidence in their ability to interpret and discuss complex topics and concepts in Japanese:

"The most rewarding part was being able to discuss and interpret such big issues and concepts in another language."

"My language confidence and flexibility have increased. This class was a great confidence builder and was really interesting."

Some students comment on a few challenges presented by the courses, such as the large amount of work involved in the half-credit course: "Homework wasn't unreasonable, but it was more time and energy than most full credit courses I've taken." Others say that they struggled from time to time with unfamiliar vocabulary and expressions related to the environment, indoctrination, medicine, laws, as well as vernacular and dialectal expressions. A few expressed feelings of being discouraged by unfamiliar genres of creative work, such as poetry, fiction, or pun, with which they had no prior experience.

From a language instructor's perspective, the careful selection of authentic, proficiency-appropriate materials in Japanese is a strenuous task, in part because environmental issues, laws, counter-measures, and recovery efforts constantly change and at a fast pace. Furthermore, the faculty's time and work demands were overwhelming, as both instructors tried to attend every session of both courses during the first year of their collaboration. On the other hand, the collaborative teaching format encouraged both instructors to acquire interdisciplinary knowledge across fields well beyond their areas of specialization, and thereby expanded their potentials to locate more resources, design and teach new courses, and organize on- and off-campus academic events, such as film screening or lectures, for winder audience across disciplines.

The FLAC model may also have a positive impact on course enrollment or curriculum revision. A few students in the REL course registered for a Japanese language course in the following academic quarters. Basic and intermediate Japanese students were encouraged to attain advanced proficiency with a clear goal to enroll in a FLAC course in the near future. Collaborative interdisciplinary teaching also drives course innovations to transform the conventional 'language-literature' learning model to an 'area-studies' model where language education is embedded with other humanities and STEM in an integrative curriculum. In the current climate of declining college language enrollments, the latter may attract more students, with fewer language or literature courses and more area-studies courses taught in English required for the major requirement (Looney & Lusin, 2018).

Despite the promising results discussed above, the lack of secured funding at our institution has made the future of our FLAC courses uncertain. Some universities are undertaking departmental or university-wide initiatives in interdisciplinary literacy curricula in the area of environmental humanities and sustainability studies (Melin, 2019). At this point, we need to build a case by providing successful interdisciplinary environmental humanities courses in as many languages as possible so that area-specific yet universally important sustainability issues can be shared and discussed beyond one country, one discipline, and one language.

Conclusion

In the spring of 2021, a new FLAC Japanese course entitled *Food and Ethics in Japan* was offered in conjunction with a new ethics course, *Religion and Ethics in Japanese Foodscapes*, as the two specialists teamed up again. Food is the most essential and fundamental element to sustain our lives as we receive the blessings of nature, animals, and plants. Therefore, food serves as a powerful analytical tool for understanding our relationship with the environment. By studying food in Japan, with particular attention to ethical implications, we can examine Japan's cultural, social, and political order. Minamata and Fukushima will be discussed again with the hope of avoiding any catastrophic food contamination in the future. Thus, environmental sustainability has become one of the central themes in our university's Japanese language and Japanese Studies programs.

Recently, a trailer was released for the film *Minamata* (Depp & Levitas, 2021), scheduled to open in early 2022. The film depicts Eugene Smith's activism, a world-renowned American war photographer who documented the devastation of the Minamata disease in the 1970s. While it will be hard to watch the physical and emotional suffering among patients and their families, I was pleasantly surprised to learn that a feature film was produced by a major film production company in the US, as this tragedy had traditionally been covered mainly in Japanese documentaries. I hope the film will make us rethink what we have done to the environment for the sake of modernization and industrialization, and those students formerly and currently enrolled in the Japanese FLAC course will watch the film and start a conversation to discuss it, in English and Japanese.

Author note: The author of this chapter would like to express their deep gratitude to Dr. Yuki Miyamoto for her guidance and collaboration in the FLAC curriculum.

Notes

1 Japanese is considered to be a Group 4 language, along with Chinese and Arabic. It takes three to four times longer to reach a specific proficiency compared to Group 1 languages, such as Spanish and French (Liskin-Gasparro, 1982).
2 The average enrollment in FLAC courses in the modern language department is three to five students.
3 The word, まち*machi*, was introduced in Lesson 4, and the kanji equivalent, 町 *machi*, in Lesson 10 of the first-year textbook, which consists of 12 lessons.

4 The author roughly translated the passage as follows, "The MACHI, which had wanted to let the people around here live a stable life, decided to ask a big company to build a big factory. Many people in the MACHI started to work at that factory."

5 Other translations of 町*machi*.

P15 町はいままでにないくらい　さわがしく、にぎやかかになった。　"The quiet <u>village</u> grew to be a bustling <u>city</u>."

P25 つぎの朝はやく、お母さんは妹をおぶって　町の病院へでかけた。しかし、なんの病気かわからずじまい。もっとくわしく調べるために、遠くの大きな町の　大きな病院へつれていかれた。"Early the next morning, mom took little sister to the doctor <u>in the village</u>, but they didn't know what was wrong. To find out, they ventured to the big hospital in a bigger <u>city</u>."

P28 妹が　かえりたがった　なつかしい町は、もう、生きものの声もしない　死んだように　ひっそりとした町に　なってしまっていた。"In the <u>village</u> little sister longed to return to, there were no voices to be heard... It was as if <u>the village</u> itself was dead."

P31 そのころ町でも、妹とおなじような病気の人が　でてきたらしい。"Meanwhile, more and more <u>villagers</u> came down with the same condition as a little sister."

6 In *Tobira* (Oka, 2009), the textbook frequently adopted in the third-year language courses at US colleges, six out of 15 chapters discuss the importance and influence of nature in Japanese geography (Chapter 1), religions (Chapter 6), history (Chapter 11), craftmanship (Chapter 12), literature (Chapter 13), and cultural values (Chapter 15).

References

Anderson, L. W., & Krathwohl, D. W. (2001). *Taxonomy for learning, teaching, and assessing: A revision of Bloom's taxonomy of educational objectives*. New York: Longman.

Ananth, P., & Lyons, L. T. (2020). *Incorporating foreign language content in a humanities course*. Abingdon: Routledge.

Atomic Age IV (2017). Website. lucian.uchicago.edu/blogs/atomicage/

Caldwell, A. M. (2001). A FLAC model for increasing enrollment in foreign language classes. *The French Review, 74*, 1125–1137.

Chikamatsu, N. (2011). Language as an essential element for human thought processes: Critical and creative thinking in content-based instruction. *Journal of Canadian Association for Japanese Language Education, 12*, 1–22.

Chikamatsu, N. (2012). Communication with community: Connecting an individual to the world through Japanese content-based instruction of Japanese American history. *Japanese Language and Literature, 46*, 171–199.

Chikamatsu, N. (2015). CCBIコース3.11地震・津波・原発: 学習者の社会参加と責任を育むために [Social responsibility and community-based learning in the post-disaster society], In S. Sato, T. Takami, U. Kamiyoshi & Y. Kumagai (Eds.), 未来を創ることばの教育をめざして内容重視の批判的言語教育 [*Language education for social future: Critical content-based instruction*] (pp. 199–220). Tokyo: Coco Publisher.

Chikamatsu, N. (2019). Collaborative teaching of a Japanese Content-based course: 3.11 and nuclear power crisis. In C. A. Melin (Ed.), *Foreign language teaching and the environment: Theory, curricula, institutional structures* (pp. 146–160). New York: The Modern Language Association of America.

Clark, B. (2016). Transdisciplinary approaches to language learning and teaching in transnational time. *L2 Journal, 8*, 3–19.

Depp, J. (Producer), & Levitas, A. (Director). (2021). *Minamata [Motion picture]*. Los Angeles: Infinitum Nihil. https://en.wikipedia.org/wiki/Minamata_(film)

Douglas, M. O. (2017). Assessing the effectiveness of content-based language instruction (CBLI) in Japanese at the college advanced level. *Japanese Language and Literature, 51*, 199–241.

García, O. (2009). Education, multilingualism, and translanguaging in the 21st century. In A. Mohanty, A. M. Panda, R. Phillipson, & T. Skutnabb-Kangas (Eds.), *Multilingual education for social justice: Globalizing the local* (pp. 128–145). Hyderabad: Orient BlackSwan.

García, O., & Kano, N. (2014). Translanguaging as process and pedagogy: Developing the English writing of Japanese Students in the US. In J. Conteh & G. Meier (Eds.), *The multilingual turn in language education: Opportunities and challenges* (pp. 258–277). Bristol: Multilingual Matters.

García, O., &Wei, L. (2014). *Translanguaging: Language, bilingualism, and education*. Basingstoke: Palgrave Pivot.

Hall, J. K. (2019). *Essentials of SLA for L2 teachers: A transdisciplinary framework*. Abingdon: Routledge.

Hanabusa, N. (2015). Joint teaching to connect Japanese language and pop culture. In S. Sato, T. Takami, U. Kamiyoshi & Y. Kumagai (Eds.), *Language education for social future: Critical content-based instruction* (pp. 247–278). Tokyo: Coco Publisher.

Hasegawa, Y. (2012). *The Routledge course in Japanese translation*. Abingdon: Routledge.

Iizuka, F. (2009). 方言と地図 [*Dialects and maps*]. Tokyo: Froebelkan.

Ishimure, M. (1964). 苦海浄土 わが水俣病 [*Paradise in the sea of sorrow: Our Minamata disease*]. Tokyo: Kodansha.

Ishimure, M. (2014). *Paradise in the sea of sorrow: Our Minamata disease*. Translated by Liva Monnet. The University of Michigan Center for Japanese Studies.

Julian. W. (2013). 東北友 [Tohoku-Tomo] [Documentary Film]. United States: Independent.

Kano, N. (2016). トランス点ランゲージングを考える:多言語使用の実態に根ざした教授法確立のために [Conceptualizing translanguaging: To consolidate pedagogy rooted in the language use of multilinguals]. *Journal of the Japanese Society for Mother Tongue, Heritage Language, and Bilingual Education, 12*, 1–22.

Kaufman, D. (2004). Constructivist issues in language learning and teaching. *Annual Review of Applied Linguistics, 24*, 303–319.

Klee, C. A. (2009). The resurgence of interest in languages across the curriculum. *The Modern Language Journal, 93*, 618–621.

Koide, Y. (2012) 原発に反対しながら研究をつづける小出裕章さんのおはなし [*Dr. Koide's lectures in anti-nuclear research*]. Tokyo: Crayon House.

Koreeda, H. (2018). 万引き家族 [*Shoplifters*] [Motion Picture]. Japan: Fuji Television.

Koyama, N. (2016). Emergence of critical reading in L2 literature seminars: Applications of discourse analysis in language pedagogy. *Japanese Language and Literature, 50*, 1–24.

Kubota, R. (2016). Critical content-based instruction in the foreign language classroom: Critical issues for implementation. In L. Cammarata (Ed.), *Content-based foreign language teaching: Curriculum and pedagogy for developing advanced thinking and literacy skills* (pp. 192–211). Abingdon: Routledge.

Kumagai, Y., & Kono, K. (2018). Collaborative curricular initiatives: Linking language and literature courses for critical and cultural literacies. *Japanese Language and Literature, 52*, 247–276.

Leung, C., & Valdes, G. (2019). Translanguaging and the transdisciplinary framework for language teaching and learning in a multilingual world. *The Modern Language Journal, 103*, 348–370.

Liskin-Gasparro, J. (1982). *ETS oral proficiency testing manual*. Princeton, NJ: Education Testing Service.

Looney, D., & Lusin, N. (2018). *Enrollments in languages other than English in United States institutions of higher education, summer 2016 and Fall 2016: Preliminary report*. New York: Modern Language Association of America.

Melin, C. A. (2019). *Foreign language teaching and the environment: Theory, curricula, institutional structures*. New York: The Modern Language Association of America.

Mieda, M. (2011). みなまたの木 [*The/A Minamata tree*]. Tokyo: Soeisha.

Ministry of Education and Science. (2013). 小学生のための放射線副読本 [*Supplementary readers' guide of radiation for elementary students*]. Tokyo: Japanese Ministry of Education and Science.

Miyazaki, H. (1984). 風の谷のナウシカ [*Nausicaä of the valley of the wind*] [Animation Film]. Tokyo, Japan: Studio Ghibli.

Oka, M. (2009). *Tobira: Gateway to Advanced Japanese learning through content and multimedia*. Tokyo: Kuroshio Shuuppan.

Okazaki, T. (2009): 言語生態学と言語教育 [*The ecology of language and language education*]. Tokyo: Bonjinsha.

Ōnuma, I. (2013). それでも僕は桃を買う [*I will still buy peaches*]. Awarded at the 33rd Middle School Student Essay Contest for Human Rights. Japanese Ministry of Justice.

Rodgers, D. M. (2015). Incidental language learning in foreign language content course. *The Modern Language Journal, 99*, 113–136.

Sato, S., Takami, T., Kamiyoshi, U., &Kumagai, Y. (2015). 未来を創ることばの教育をめざして内容重視の批判的言語教育 [*Language education for social future: Critical content-based instruction*]. Tokyo: Coco Publishing.

Someya, Y. (2010). 大学における翻訳教育の位置づけとその目標 [The teaching of translation in the context of college education: Its purpose and rationale]. *Journal of Kansai University Faculty of Foreign Language Studies, 3*, 173–102.

Specter, D. (2011). いつも心にクールギャグを [*Cool gags in your heart*]. Tokyo: Gentosha.

Stryker, S. B., & Leaver, B. L. (1997). *Content-based instruction in foreign language education: Models and methods*. Washington, DC: Georgetown University Press.

The Douglas Fir Group. (2016). A transdisciplinary framework for SLA in a multilingual world. *The Modern Language Journal, 100*, 19–47.

Walker, B. L. (2010). *Toxic Archipelago: A history of industrial disease in Japan*. Seattle: University of Washington Press.

Conclusion

María J. de la Fuente

In the post-COVID world, we are already facing a rapid transformation of universities that could be taken as an opportunity to rethink their educational mission and their relevance and contribution to sustainable development and the well-being of society. Such a transformation should result in educational goals grounded on the principles of inclusion; equity; human rights; social justice; peace; ethnic, cultural, and linguistic diversity; and education for sustainable development.

Foreign languages and cultures encompass the values, beliefs, knowledge, arts, traditions, and ways of life through which peoples and communities express and develop. For example, the study and understanding of foreign cultures, including minoritized cultures, is crucial for sustainable development goals (SDGs) related to consumption and production. Another SDG—promoting inclusive and peaceful societies—cannot be achieved without the study and knowledge of foreign languages and intercultural communication; not if we want the most vulnerable groups in society—migrants, refugees, indigenous communities—to have access to sustainable development. These are just two of the multiple ways in which knowledge of foreign languages and cultures is a required transversal dimension to achieve the goals of any humanistic sustainable development agenda.

In the post-COVID context, we face some harsh realities: Growing inequality, unemployment hitting our college graduates hard, and our curricula that do not fit the changing job markets... If there ever was a time to reevaluate what we do, it is now. A student-centered curriculum cannot ignore the post-COVID world. College foreign language curricula that educate for sustainable development can (a) make students multicompetent language users who can engage in intercultural dialogue, (b) give them relevant interdisciplinary knowledge, and (c) provide them with transversal, high-level cognitive skills (problem-solving, critical thinking) and transferable, life-long skills (communication, teamwork, conflict resolution).

The foreign language educators who contributed to this volume already shared, in belief and practice, the idea that education in foreign languages and cultures needed to be connected to inclusion, equity, social justice, cultural and linguistic diversity, and other relevant areas. What they did in these projects was to take foreign language curricular content and place it under a

DOI: 10.4324/9781003080183-17

larger, global educational framework: Sustainable development, which makes it more relevant for their students, for their own disciplines, and for the university and the community at large. They designed great learning experiences, many of which happened amid a pandemic, for the students' benefit.

Finally, from this experience, we learned that educating for sustainability in foreign languages can happen everywhere: We now look forward to going back to our physical classroom where we can implement learner-centered approaches based on dialogue and social interaction; to our distance-learning classroom where we can engage in intercultural dialogue and even translanguaging; and to our street classroom, our museum classroom, or our strawberry field classroom, where we will engage in project-based or community-based language learning.

Index

Note: **Bold** page numbers refer to tables, *italic* page numbers refer to figures and page numbers followed by "n" denote endnotes.